AN INTRODUCTION TO TRIGONOMETRY AND ITS APPLICATIONS

Authored By

Vasile Postolică

Romanian Academy of Scientists,

Vasile Alecsandri University of Bacău,

Bacău, Calea Mărăşeşti

Romania

DEDICATION

THANK GOODNESS

CONTETNS

Foreword

To many generations trigonometry was one of the mathematical subjects taught in school, along with arithmetic, algebra, geometry, and analysis. For some time now, due to the need to diminish the syllabus, trigonometry stops being considered a distinct subject with its own handbook, and is included in the syllabus for algebra, geometry and analysis. Under these circumstances, the unitary nature of trigonometry is sacrificed. But there are still enough pupils interested in mathematics, enough graduates and students, not to mention the math teachers, the numerous engineers and economists that are still fond of elementary, school mathematics, but who would like to reconsider it from a superior point of view, that is from the perspective of mathematic knowledge accumulated in time. This book, written rigorously, but passionately, has a perspective that is equally based on the ensemblist structure of modern mathematics, on the elements of real analysis and the axiomatic fundaments of geometry, and it reaches climax with the applications that imply analytic algebraic and geometric abilities. From the historical presentation of mathematics given in the book the reader can, for example, learn the surprising fact that, from a chronologically point of view, plane trigonometry was preceded by spherical trigonometry, necessary in astronomy, although the latter is more complex than the former. The author of this book, Professor Vasile Postolică, teaching at "Vasile Alecsandri" University of Bacău, is a mathematician that has convincingly asserted himself in research, with results published in exacting publications, results that have become a part of the international circuit. His vast experience in school and university teaching made possible the development of this successful experiment, experienced by few professors: the re-writing of a chapter of school mathematics from a superior point of view. This adventurous path has been taken by mathematicians like Felix Klein and Jacques Hadamard, hence this is a very serious project. We congratulate Professor Vasile Postolică for this success. At the same time, supposing that he will continue this adventure, we suggest that he broaden the perspective by introducing elements of trigonometric series and Fourier series, that are, probably, the most important use of the trigonometric functions, having a particular relevance in physics. Some elements of spherical trigonometry would be also timely. The present book deserves to be successful in book stores and justifies every one of our praises to its author.

Academician Solomon Marcus,
Piatra Neamţ,
România

PREFACE

Trigonometry appeared as a part of medieval geometry, at first the spherical component required by astronomy, after that plane trigonometry was developed. In 1260 the most extensive work in this field was created " The Treatise of the Complete Quadrilateral", written by Näsr-ad-Din, followed by the first synthesis of trigonometric knowledge of the 15th century elaborated by Regiomontanus. By studying and developing applied mathematics, the Indians have introduced the sine function as association of any circle arc having half the length of the under tightened cord by the arc's double, the term "sine" and some interpolator methods of calculating the values of this function were given by Peuerbach, and the natural connections between algebra, mathematical analysis and trigonometry appeared in the 18th century. Using the major elements of the Basis of Mathematics, the book briefly, but rigorously presents in the first chapter sets, relations, and the algebraic-topologic structure for R and $\overline{R}$, properties of real functions of real argument in the second chapter, and the concept of vector in R^3, described in the third chapter. The book presents itself as an original and synthetic approach that allows a quick learning of the fundamental trigonometric functions and the corresponding inversive restrictions thoroughly analyzed in the fourth chapter, followed by the suggestive applications in the last chapter, without any further consultation of some additional bibliography. The theoretical considerations and the applicative issues are significant proving to be very useful for everyday classes, different exams and the methodic-scientific improvement. Thus, the book addresses anyone who is interested not only in properly knowing trigonometry, but also in the major elements regarding one of the essence of mathematics. It is the result of the unrests of a teenager fascinated by mathematics and of the responsibilities of the teacher he has come to be and it is also of interest to: advanced high school students, undergraduates, and mathematics teachers. The author shows great responsiveness towards any suggestion meant to improve the contents and the presentation, with thanks everyone who has made possible the publishing of this book.

ACKNOWLEDGEMENT
None Declare
CONFLICT OF INTEREST
None Declare

Vasile Postolică,
Romanian Academy of Scientists,
"Vasile Alecsandri" University of Bacău,
România,
Email: vpostolica@ambra.ro

CHAPTER 1

Sets. Relations. R and $\overline{R}$

Abstract: In this chapter we will briefly present the fundamental terminology of mathematics regarding notions such as set and, respectively, relation, mentioning the algebraic-topological structures for R and $\overline{R}$ sets.

Keywords: Axiom, geometry, cardinal, relation, function, real number, extension, completeness axiom, Cantor- Dedekind, topology, implication.

1.1. INTRODUCTION

Axiomatization and formalization are specific for nowadays science and have their roots in Greek antiquity. Thus the history of mathematics records as the first idea of axiomatic organization of science of that period as belonging to Parmenides and the elaboration of the first geometric theory under ancient axiomatic form to Euclid (approx. 365-300 B.C.). He divided the elements that were accepted without definition and/or demonstration from those that could have been given by definition, respectively by deduction, stating also some rules of deduction. Euclid accepted the following distinction between axiom and postulate: the axiom was accepted as having intuitive evidence, with no possible negation, and the postulate was a certain empiric sentence, that had an unchallenged truth, meaning that the demonstration, possibly followed by a negation that was admitted without having always a possible justification could not be given yet. The deeply philosophic nature of Greek mathematics allowed that both Euclid's Postulate and the next negations to increase the quality of the mathematical thinking and civilization.

(P_1) - *There is no parallel to a straight line passing through any exterior point of that line* (this subsequently generated the geometry of Bernhard Friedrich Riemann (1826-1866)).

(P_2) - *Through any exterior point of any straight line there are at least two parallels to that line* (in 1825-1826, the geometry of Bolyai Janos (1802-1860) and Lobacevscki Nikolai Ivanovici (1792-1856) was based on this postulate), both of them determined the appearance of "classic" non-Euclidean geometries, after the appearance of the first two-dimensional non-Euclidean geometry developed by Menelaus from Alexandria (approx. 98 A.D.) in the first book of the treatise that had three volumes, entitled "The Spheres".

The systematic presentations of geometry entitled "The Elements" of Hippocrates of Hios (the 5th century B.C.) are considered to be primordial (they were probably out-shone by Euclid's "The Elements", that appeared approximately in the 3rd century B.C.). In the latter, the geometry is presented under the form of a logic system so well organized that nothing of principle could be added for more than two millennia, that is till the appearance of the unchallenged non-Euclidean geometries (Euclid, by synthesizing the results of the further development, presented the geometry as a autonomous, logically organized, theoretical science, thus laying the fundaments- within the limitations of the age- of the Axiomatic Method, having a fundamental role both in modern Mathematics and in the other sciences).

These geometries that marked the natural evolution from ancient geometry, that was rooted in Pharaohs' Egypt and was subsequently considered a science by the Greeks (Thales of Millet (approx.640-548 B.C.), Pythagoras (580-500 B.C.), Plato (428-348 B.C.), Archimedes (287-212 B.C.), Apollonius of Perg (262-200 B.C.), Euclid(the 3rd century B.C.), Nicomedes(the 3rd- 2nd century B.C.), Diocles(the 2nd century B.C.), etc.) to the modern geometry showed that the concept of "axiom" in the Euclidean sense wasn't correct at least from the logical point of view.

Thus, the Euclidean geometries, in which the justification of Euclid's Postulate (the famous *Axiom of parallels*) is accepted, associated with the philosophic thinking allowed the acknowledgement of the concept of axiom not as a truth with intuitive evidence, but as something abstract, that allows the coherent organization of a theory. Starting from the axioms, the results included in the theorems and the corresponding consequences are established by using the customary deduction rules. This method of constructing a theory following the mentioned requirements is called the *hypothetic-deductive method.*

The axiomatized theories are hypothetic-deductive theories in which the initial terms and the primary sentences are explicitly presented. These theories have had two stages of development, according to the conception on axiom.

Within the theories belonging to the first stage the objects of the subject matter that is axiomatized become known before the axioms; the axioms are properties of the objects, being an evidence, and the subject matters based on experiments are rooted in those mentioned above (for example, Euclid's Geometric Theory).

In the second stage, the formal axiomatic is characterized by the fact that the axioms precede the system of objects to which they refer, thus leading to the *formal deductive theories*

The logic requirements of any axiomatic theory are the following:

- *non-contradiction;*

- *independence;*

- *completeness;*

- *categoricalness* (an axiomatic theory is *categorical* if any two illustrative models are isomorphous).

We will mention some of the notorious systems of axioms:

- The first attempt to axiomatize geometry made by Euclid in the paper entitled " The Elements", that has 13 books;

- The system of axioms belonging to David Hilbert (1862-1943), the first complete axiomatic system elaborated in 1899 added to geometry (a slightly different form was indicated by Birkhoff George David (1884-1944)).

- The system of axioms of Giuseppe Peano (1858-1932) regarding the axiomatic introduction of the natural numbers set established in 1889-1895, with the corresponding variants of definition for the real numbers. It is fit to mention that the first axiomatic construction of the natural number set was anticipated by Richard Dedekind (1831-1916) and later mentioned by Peano.

- The axiomatic system of the set theory elaborated in Italy by the German mathematician Ernst Zermelo (1871-1953) in 1908 and improved in Germany and Scandinavia in 1919-1921 by Abraham Fraenkel (1891-1965) and Thoralf Skolem (1887-1963).

1.2. SETS

Georg Cantor (1845-1918) presented in Zurich, in 1897, the notion of *set* as *"a unitary whole, with distinct elements, in which the order of element display is not important".*

The mathematician David Hilbert considered that Cantor's contribution was *"the most comprehensive result of mathematic thinking, one of the most beautiful accomplishments of human thinking in the field of intelligible."* But not any class of objects is a set.

Thus, if we consider the *Ens* family as being of all the sets and if we admit that this is a set, then $T = \{x \in Ens : x \notin x\}$ would also be a set, situation in which it can't be decided if $T \in T$ or $T \notin T$. This paradox (this antinomy) of Bertrand Russell (1872-1970), which appeared in 1899 together with others, has shown that the central elements of Cantor's Theory had to belong to a system of axioms that met all the fundamental requirements: non-contradiction, independence, completeness and categoricalness that are typical of any system of axioms on which a science is usually based.

We mention that, generally, the antinomies (paradoxes) can't be removed so easily; consequently we can't say that the problem of the appearance of antinomies in mathematics can be considered solved. But, due to the fact that they appear in special situations that do not immediately affect the Foundations of Mathematics or the current Applications of Mathematics, the paradoxes are, for the moment, an undiscovered chapter of the potential of human thinking.

Mathematicians E. Zermelo and A. Fraenkel have suggested a method of solving the problem of antinomies with the help of an adequate system of axioms (see, for example, [1]):

(Z1) The axiom of determination

> If A and B are arbitrary non-empty sets, any element of A belongs to B and reciprocally, then $A=B$.
>
> Whenever A and B are both empty sets one considers also that $A=B$.

(Z2) The axiom of elementary sets

> (1) *There exist empty sets; any of these sets being denoted by Ø.*
>
> (2) *If a is an arbitrary object, then there exists the set {a} in which a is the only element.*
>
> (3) *If a and b are different "objects", then there exists the set {a,b} in which a and b are the only elements.*

(Z3) The axiom of the base

> *Any non-empty set X has at least an element y so that x and y have no common elements; if p(x) is any property or an arbitrary ensemble of properties regarding the elements x of every set A, then there is a set T that includes those elements $a \in A$ for which p(a) is logically valid and that doesn't include other elements.*

(Z4) The axiom of the subsets

> *For any set A there is a set* P *(A) that contains exactly the subsets of A.*

(Z5) The axiom of reunion

> *For any set A of sets there exists a set F that has only the elements of the sets included in A.*

(Z6) The axiom of choice (this is considered to be the most important and the most debated upon by the mathematicians, philosophers etc.)

For any set H of non-empty sets, mutually disjoint, there exists a set that has exactly one element of every set included in H.

(Z7) The axiom of the infinite

There is a set M that meets the following requirements:

> *(1) $\varnothing$ is an element of M;*

> *(2) If a is in M, then {a} is as well in M.*

As we have already anticipated, with respect to Zermelo's Axiomatic, the class of all sets is not a set.

Also, in this context, it is clear the validity of

(Z8) The principle of abstraction

For any non-empty set X and any predicate $P_X(\cdot)$, there exists a set that includes only the elements $y \in X$ for which $P_X(y)$ is a true sentence.

Before presenting equivalent sentences to the axiom of choice, we shall exhibit some useful preliminary considerations.

Let $(X, \leq)$ be an arbitrary non-empty, pre-ordered set.

Definition 1. *An element $a_0 \in X$ ($b_0 \in X$, respectively) is called the first or the lowest element, with respect to the pre-order relation "$\leq$" (the last or the highest element, respectively) if $a_0 \leq x$, $\forall x \in X$ ($x \leq b_0$ $\forall x \in X$, respectively).*

Definition 2. *$x_0 \in X$ is called minimal (maximal, respectively)*

if $x \leq x_0$, $x \in X \Rightarrow x = x_0$, ($x_0 \leq x, x \in X \Rightarrow x = x_0$, respectively).

A non-void set is called *well ordered* whenever it is fully ordered and any non-empty subset allows a first element.

Every fully ordered subset of any ordered set is called a *chain*. If in an arbitrary ordered set any chain is upper ordered bounded, then the set is called an *inductive ordered set*.

By the *selective function* attached to any non-empty set *A* we understand any function of the type: $f : P_0(A) \to A$ with

$f(B) \in B, \forall B \in P_0(A)$, where $P_0(A)$ designates the family of all non-void subsets for the set A.

Theorem 1. *The next statements are equivalent to each other:*

(1)   The axiom of choice;

(2) **Zorn's lemma**: Any inductive ordered set has a maximal element;

(3) **The principle of well ordered (Zermelo)**: Any non-empty set can be well ordered;

(4) **Kuratowski's lemma**: Any chain of every partially ordered set is a part of a maximal chain (with respect to the inclusion relation).

(5) **Zermelo's axiom**: For any non-emty set there exists a selective function.

Thus, any set that doesn't have elements will be called *empty (void)* and will be denoted with $\varnothing$. If A and B are sets so that any element of A is found in B, then A is called *a subset* (a part) of the set B and we write $A \subseteq B$. It is considered that $\varnothing \subseteq A$ for any set A. The subsets that are different of $\varnothing$ and A are called *proper*, and $\varnothing$ and A are called *improper parts* of any non-void set A.

If A and B are sets that meet the requirement $A \subseteq B \subseteq A$, then $A=B$. Under any other circumstance, it is written $A \neq B$. By $A \subset B$ it is understood $A \subseteq B$ and $A \neq B$.

The *reunion* of two arbitrary sets A and B is defined by

$$A \cup B = \{x : x \in A \text{ or } x \in B\},$$

their *intersection* is given by

$$A \cap B = \{x : x \in A \text{ and } x \in B\},$$

their *difference* being defined by

$$A \setminus B = \{x : x \in A \text{ and } x \notin B\}.$$

If $B \subseteq A$, then $A \backslash B$ is called *the complementary set* of the set B in relation (with respect) to set A. We want to mention that the difference (the complementariness) of sets must not be mistaken with the *algebraic difference* of two non-empty arbitrary sets from any additive group which is defined by: $A - B = \{a\text{-}b: a \in A \text{ and } b \in B\}$.

The *Cartesian product* of the sets A and B is described by $A \times B = \{(a,b): a \in A \text{ and } b \in B\}$, considering that $A \times B = \varnothing$ whenever $A = \varnothing$ or $B = \varnothing$.

Similarly, if $\{A_i\}_{i \in I}$ is an arbitrary family of sets, then their *reunion* is

$$\bigcup_{i \in I} A_i = \{x: \exists i \in I \text{ with } x \in A_i\},$$

the *intersection* is defined by

$$\bigcap_{i \in I} A_i = \{x: x \in A_i, \forall i \in I\},$$

and the *product* by

$$\prod_{i \in I} A_i = \{f: I \to \bigcup_{i \in I} A_i \,/ f \text{ is a } function \text{ with } f(i) \in A_i, \forall i \in I\}.$$

When the set of indices I has $n \in N^*$ elements, then the notations $\bigcup_{i=1}^{n} A_i, \bigcap_{i=1}^{n} A_i$ are used correspondingly, and the product $\prod_{i=1}^{n} A_i$ coincides with the set

$$\{(a_1, a_2, ..., a_n): a_i \in A_i, i = \overline{1,n}\}.$$

1.3. RELATIONS

Let A and B be arbitrary non-void sets. Any triplet $\left(A, B, \rho \subseteq A \times B\right)$ is called a *binary relation* defined by the sets A and B.

Instead of using the notation $(a,b) \in \rho$ $(a \in A \text{ and } b \in B)$ it is written $a \, \rho \, b$ and it is read "*a* is in the relation ρ with *b*". The sets

$$dom(\rho) = \{a \in A: \exists\, b \in B \text{ with } a\,\rho\,b\},$$

$$codom(\rho) = \{b \in B: \exists\, a \in A \text{ with } a\,\rho\,b\}$$

are called the *domain* and, respectively, the *co-domain* of the relation ρ, and the inverse relation ρ^{-1} defined by the sets A and B is given by $b\rho^{-1}a \Leftrightarrow a\rho b$ and it is called the *inversion* of the relation ρ.

Whenever the sets A and B coincide, any binary relation defined on the the same set is called *homogeneous.*

Any homogeneous relation $(A,A,\rho\subseteq A^2)$, which is reflexive $(a\rho a, \forall a \in A)$, symmetrical $(\forall a_1,a_2 \in A$ with $a_1\rho a_2 \Rightarrow a_2\rho a_1)$ and transitive $(\forall a_1,a_2, a_3 \in A$ with $a_1\,\rho\,a_2$ and $a_2\,\rho\,a_3 \Rightarrow a_1\,\rho\,a_3)$, is called *relation of equivalence.*

In all the situations such as these, to every element $a \in A$ can be added the corresponding *class of equivalence* $C(a)=\{a' \in A: a'\,\rho\,a\}$.

It is clear that $C(a) \neq \varnothing$, $\forall a \in A$, $C(a_1) \cap C(a_2)=\varnothing$, if a_1 and a_2 are not elements of the same class and $\bigcup_{a \in A} C(a) = A$, that is, the family $\{C(a): a \in A\}$, also called the *factor set* of A by ρ and marked by A/ρ, makes a *partition* of the set A.

Examples. The equality relation of the sets, the equality relation of the functions, the relations of parallelism for straight lines and planes, respectively, in general, the relation of congruence of sets in any usual Euclidean space. If one considers that two arbitrary non-empty sets A and B are *equipotent*, in notation $A \approx B$ whenever there exists at least a bijective function $f : A \to B$, then the relation of equipotence is also a relation of equivalence and any class of equivalence designates a cardinal. Thus,

$$card(A) = \{B \in Ens : B \approx A\}, \forall A \in Ens \ .$$ Whenever for a set $A \in Ens$ we have $A \approx N$, one says that A is *a countable* set (like *Q*) .Any finite or countable set is called *at most countable.*

Every binary homogeneous relation which is reflexive and transitive is called a *pre-order relation.*

Any of pre-order relation $\rho \parallel$ on every non-empty set A which is anti-symmetrical $\left(\forall a_1, a_2 \in A \right.$ with $a_1\rho a_2$ and $a_2\rho a_1 \Rightarrow a_1=a_2$ is called a *order relation*. Thus, for example, the relation of strict inclusion of sets is a   pre-order relation, and that of inclusion is an order relation.

By *function* (functional relation), defined on empty set A, called *set of definition*, with values in an arbitrary non-empty set B, called *set in which the function has values*, is meant any univocal binary relation $\left(A, B, f \subseteq A \times B\right)$ $\left(\forall a_1, a_2, a_3 \in A \right.$ with $a_1 f a_2$ and $a_1 f a_3 \Rightarrow a_2 = a_3 \left.\right)$.

The binary relations of this type are marked by $f : A \to B$. A function f of this type is called *injective* if the inversion f^{-1} of the relation of origin is univocal and surjective if *dom(f^{-1})=B*. Any injective and surjective function is called *bijective function*. It is evident that a function $f : A \to B$ is injective if and only if for any $a_1, a_2 \in A$ with $a_1 \neq a_2 \Rightarrow f(a_1) \neq f(a_2)$ or, equivalently, for any $u, v \in A$ for which $f(u) = f(v)$ the result is only $u = v$. Clearly, if $f : A \to B$, is any functional relation, f^{-1} is a function if and only if f is injective

Two functions $f : A \to B$ and $g : D \to E$ are   called *equal*, in notation $f=g$, if $A=D$, $B=E$ and $f(x) = g(x)$, $\forall x \in A$.

For any non-empty set A, the *identical* application, defined by $1_A(a)=a$, $\forall a \in A$, will be marked by $1_A : A \to A$.

If A_1, A_2, A_3 are non-void sets, and if $f_1 : A_1 \to A_2$ and $f_2 : A_2 \to A_3$ are functions, then the *composed* function $f_2 \circ f_1 : A_1 \to A_3$ can be defined by $(f_2 \circ f_1)(x) = f_2[f_1(x)], \forall x \in A_1$. A function $f : A \to B$ allows an inversion if and only if there is a function $g : B \to A$ so that $g \circ f = 1_A$ and $f \circ g = 1_B$. If A_1, A_2, A_3 are arbitrary non-empty sets and $f : A_1 \to A_2, g : A_2 \to A_3$ are functions, then the function $g \circ f$ is injective, whenever f and g are injective, and surjective, when f and g are surjective.

1.4. THE REAL NUMBERS SET

Due to the fact that the fundamental concept of *real number* still proves to be insufficiently known and, especially, assimilated we will present here a synthesis of the modern ways of axiomatic introduction, respectively, of construction regarding the *real number set*, having in view also the historic evolution of this notion. It appears like the best illustration as a topological space and an ordered topological vector space, respectively (see, for instance, [2], [3]).

In 1889-1895, Giuseppe Peano developed the mentioned system of axioms for the natural number set *N:*

> (P1) *0 is a natural number.*
>
> (P2) *The successor of any natural number is also*
>
> *a natural number.*
>
> (P3) *There are no distinct natural numbers having*
>
> *the same successor.*
>
> (P4) *0 is not the successor of any natural numbers.*
>
> (P5) *If a set of natural numbers includes the*
>
> *number zero and with every natural number*
>
> *includes also the corresponding successor, the A=N.*
>
> The successor of every number $a \in N$ is marked with a'.

The immediate consequences

1) $a,b \in N; a = b \Rightarrow a' = b'$;

2) If $a,b \in N$ then $a \neq b \Leftrightarrow a' \neq b'$;

3) There exists an unique function $\varphi : N \times N \to N$ with the relations $\varphi(m,0) = m, \forall m \in N$ and $\varphi(m,n') = [\varphi(m,n)]'$ that, in additive notation, become $m + 0 = m, \forall m \in N$ and $m + n' = (m+n)', \forall m,n \in N$, respectively.

4) There exists an unique function $\psi : N \times N \to N$ so that $\psi(m,0) = 0, \forall m \in N$ and $\psi(m,n') = \varphi[\psi(m,n),m], \forall m,n \in N$.

If we write $\psi(m,n) = m \cdot n$ then the characteristic properties for the function ψ are written $m \cdot 0 = 0, \forall m \in N$,

$m \cdot (n+1) = m \cdot n + m$ and   the   validity   of   the   following equalities is also checked:

5)   *0+n=n, $\forall n \in N$;   m'+n=(m+n)', $\forall m,n \in N$;   m+n=n+m, $\forall m,n \in N$; (m+n)+p=m+(n+p), $\forall m,n,p \in N$; n'=n+1, $\forall n \in N$ and 1=0';*

$$m{+}n{=}0 \Leftrightarrow \begin{cases} m = 0 \\ n = 0 \end{cases}; \ m{+}n{=}1 \Leftrightarrow \begin{cases} m = 0 \\ n = 1 \end{cases} \text{or} \begin{cases} n = 0 \\ m = 1 \end{cases};$$

$$0 \cdot m = 0, \forall m \in N \ ; m' \cdot n = mn + n, \forall m, n \in N \ ; \ n \cdot m = m \cdot n, \forall m, n \in N \ ;$$

$$(m \cdot n) \cdot p = m \cdot (n \cdot p), \forall m, n, p \in N \ ; n \cdot 1 = n, \forall n \in N$$

$$n \cdot (m + p) = m \cdot n + n \cdot p, \forall m, n, p \in N \ ;$$

$$\left. \begin{matrix} n \cdot p = n \cdot m \\ n \neq 0 \end{matrix} \right\} \Rightarrow m = p; n \cdot m = 0 \Leftrightarrow n = 0 \ \textit{or m=0}; \ n \cdot m = 1 \Leftrightarrow n = m = 1.$$

Presenting these considerations the purpose mentioned at the beginning of this paragraph can be achieved in the following way:

A.   *N={0,1,2,3,...,n,...}, N*=N\{0}={1,2,3,...,n,...},*

Z=(-N)$\cup$(N)={...-n,...,-3,-2,-1,0,1,2,3,...,n,...},

Q={p/q:p,q$\in$Z,q$\neq$0} is the set of all the decimal, periodical numbers with the period different of (9),

R\Q is the set of all decimal numbers, having an infinity of decimals, in which no decimal or group of decimals are repeated periodically, *R=Q$\cup$R\Q* is the set of real numbers having the algebraic, order and topologic structure indicated by the axiomatic construction and by the following considerations.

We mention that the natural number set can also be introduced by using the cardinals defined in the 1.2 section:

$$0 = card(\phi), 1 = card\{\phi\}, 2 = card\{\phi, \{\phi\}\}, \text{ and so on.}$$

The numbers included in the set *N* were introduced in our civilization by the Indian culture, being accepted after a millennium of uncertainty. The hardest to identify were the *irrational numbers* which, although they were considered by the Pythagoreans as "lengths of segments", a status that was accepted also by the antic Greeks, have produced a *giant cognitive leap* for the human knowledge. In this sense, the poet-mathematician Dan Barbilian (Ion Barbu) (1895-1961) stated:

"The crisis of the Greek civilization was determined by the impossibility of conceiving the irrational number".

The communion between the *Arab, Babylonian, Egyptian, Greek and Indian* civilizations have imposed these numbers as well, the following being the complex numbers introduced through geometric representations by Carl Friedrich Gauss (1777-1855), and in 1837 by Rowan Hamilton(1805-1865) as ordered pairs of real numbers from every Euclidean plane, being later grounded, but only after Georg Cantor, Augustin

Louis Cauchy (1789-1857), Richard Dedekind and Karl Weierstrass (1815-1897) have irrevocably specified the *real number*.

B. Considering the set N axiomatically constructed by Peano, in $N \times N$ the following relation of equivalence will be introduced:

$$(a,b) \sim (c,d) \Leftrightarrow a+d = b+c$$

We will write $Z = N \times N / \sim$, thus $Z = \{\overline{(a,b)} : a,b \in N\}$ where

$$\overline{(a,b)} = \{(c,d) \in N \times N : (c,d) \sim (a,b)\}$$

and we define $+, \cdot : Z \times Z \to Z$ by

$$\overline{(a_1,b_1)} + \overline{(a_2,b_2)} = \overline{(a_1+a_2, b_1+b_2)},$$

$$\overline{(a_1,b_1)} \cdot \overline{(a_2,b_2)} = \overline{(a_1 a_2 + b_1 b_2, a_1 b_2 + a_2 b_1)}, \ \forall \overline{(a_1,b_1)}, \overline{(a_2,b_2)} \in Z$$

It is also easy to demonstrate that $(Z,+,\cdot)$ is a commutative unitary ring, with no dividers of zero (integrity domain).

Let $Z^* = Z \setminus \{0\}$ and in $Z \times Z^* = \{(a,b): a \in Z, b \in Z^*\}$, let φ be the equivalence relation introduced by: $(a,b)\varphi(c,d) \Leftrightarrow ad = bc$.

By definition, the set $Q = Z \times Z^* / \varphi$ of all the classes of equivalence in relation with φ is called the *real numbers set.*

R can be defined at this point as the set of all the classes of equivalence corresponding to the Cauchy rational number sequences in the next relation denoted by "~" : if (r_n), (r_n') are two Cauchy rational number sequences, then

$$(r_n) \sim (r_n') \Leftrightarrow r_n - r_n' \to 0.$$

This way of conceiving the notion of real number with the precedent description still is nowadays the most appropriate method in relation to the human possibilities regarding the decimal representations of the real numbers, being characteristic to the iterative algorithmic thinking.

C. Richard Dedekind's arithmetization

It is supposed that the set *N* is constructed following the below enlisted axioms, which generated those included in Peano's system of axioms:

(D1) $0 \neq x+1, \forall x \in N$:

(D2) $x+1 = y+1 \Leftrightarrow x = y$ *when* $x,y \in N$;

(D3) *if* $\varnothing \neq M \subseteq N$, the M has the lowest

element compared to the "$\leq$" relation, given by

(D4) $x \leq y(x,y \in N) \Leftrightarrow \exists z \in N : x+z = y$:

(D5) *the operations written additively,*

respectively, multiplicatively meet the relations:

$$x+(y+1)=(x+y)+1, x+0=x, x\cdot(y+1)=x\cdot y+x, x\cdot 0=0, \forall x,y\in N.$$

Q is defined as a fully ordered field made up of the set of classes of equivalence $\overline{(a,b,c)}$ of the triplets $(a,b,c)\in N^3$ with $c\neq 0$ by the following relation of equivalence:

$$(a,b,c)\equiv(a',b',c') \Leftrightarrow ac'+b'c=a'c+bc',$$

and the usual operations are given by:

$$\overline{(a,b,c)}+\overline{(a',b',c')}=\overline{(ac'+a'c,bc'+b'c,cc')}$$

$$\overline{(a,b,c)}\cdot\overline{(a',b',c')}=\overline{(aa'+bb',ab'+ba',cc')}$$

In this context $0=\overline{(0,0,1)}, 1=\overline{(1,0,1)}$, the relation of order

is given by

$$\overline{(a,b,c)}\leq\overline{(a',b',c')} \Leftrightarrow ac'+b'c\leq a'c+bc'$$

and

$$\frac{a-b}{c}=\{(a',b',c')\in N^2\times N^*:(a,b,c)\equiv(a',b',c')\}=\overline{(a,b,c)},$$

and $Z=\{(a,b,1):a,b\in N\}$.

Any non-empty set $S\subset Q, S\neq Q$ that doesn't include the highest element with respect to the relation "$\leq$" and with the property: every time $x\in S$ and $y\leq x$ the result is $y\in S$ is called a *Dedekind cut in Q*. Under these circumstances, the real number set R is *the completely ordered field of all Dedekind cuts* $S\subset Q$, and the customary operations and the relation of order are defined in the following manner:

$$S+T=\{x+y:x\in S,y\in T\}$$
$$S\cdot T=\{x\cdot y:x\in S,y\in T\}, \forall S,T\in R$$
$$0=\{x\in Q:x<0\}$$
$$1=\{x\in Q:x<1\}$$

and if $S,T\in R$, then $S\leq T \Leftrightarrow S\subseteq T$.

D. *By the term real number set is understood any commutative field* $\left(R,+;:R^2\to R,<\right)$ *fully ordered by a relation* $"\le"$, *thus it has the following properties:*

d_1) $x+\left(y+z\right)=\left(x+y\right)+z,\forall x,y,z\in R$ (+ - associative);

d_2) $\exists 0\in R,\forall x\in R:x+0=0+x=x$ (there is a neuter in relation to "+" marked with 0);

d_3) $\forall x,\exists\left(-x\right)\in R:x+\left(-x\right)=\left(-x\right)+x=0$ (any element $x\in R$ allows an opposite $-x\in R$);

d_4) $x+y=y+x,\forall x,y\in R$ (+ - commutative);

d_5) $x\cdot\left(y\cdot z\right)=\left(x\cdot y\right)\cdot z,\forall x,y,z\in R$ ($\cdot$ - associative);

d_6) $\exists 1\in R,\forall x\in R:x\cdot 1=1\cdot x=x$ (there is a neuter element in relation to "$\cdot$" marked with 1;

d_7) $\forall x\in R\setminus\{0\},\exists x^{-1}\in R:x\cdot x^{-1}=x^{-1}\cdot x=1$ (any element $x\in R\setminus\{0\}$ allows an inversion $x^{-1}\in R$);

d_8) $x\cdot y=y\cdot x,\forall x,y\in R$ ($\cdot$ - commutative);

d_9) $x\left(y+z\right)=x\cdot y+x\cdot z,\forall x,y,z\in R$ ("$\cdot$"is distributive compared to "+");

d_{10}) $\forall x,y,z\in R$ with $x<y$ and $y<z\Rightarrow x<z$ ("$<$"is transitive);

d_{11}) $\forall x,y\in R\Rightarrow x<y$ or $x=y$ or $y<x$ (the property of trichotomy of the order relation "$<$"), by $x\le y$ it is meant $x<y$ or $x=y$;

d_{12}) $\forall x,y\in R$ with $x<y\Rightarrow x+z<y+z,\forall z\in R$ (the compatibility of "<" with the "+" relation);

$d_{13})$ $\forall x, y \in R$ with $x < y$ and $\forall z \in R$ with $0 < z \Rightarrow x \cdot z < y \cdot z$ (the compatibility of "<" in relation to the multiplication using positive real numbers)

in which the **Axiom of the upper boundedness (the Axiom of completeness Cantor-Dedekind)** or, its equivalent form, the **Axiom of the lower boundedness** are checked and the both of them are indicated by the following considerations. The real number set is identical by appropriate isomorphisms with any fully ordered commutative field that meets the **Axiom of completeness Cantor-Dedekind.**

The considerations below are devoted to some remarkable properties which play a central role in the understanding of the corresponding algebraical an topological structure for R.

As it is well known, the classical Mathematical Analysis was developed on any such as this fully ordered commutative field.

Definition 1. *A non-empty real numbers set A is called upper bounded if there exists $\alpha \in R$ so that $a \leq \alpha$ for any $a \in A$.*

The Axiom of the Upper Boundedness (The Axiom

of Completeness Cantor-Dedekind).

For any non-empty upper bounded real numbers set A there exists the lowest increaser in R called the superior margin of the set A and it is marked with sup (A).

If sup $(A) \in A$, then it will be written sup(A) = max(A).

The following statements are equivalent with each other:

(i) $M \in R$ and $M = \sup(A)$;

(ii) $\begin{cases} a \leq M, \forall a \in A; \\ \forall M_1 \in R \text{ with } a \leq M_1, \forall a \in A \Rightarrow M \leq M_1; \end{cases}$

(iii) $\begin{cases} a \leq M, \forall a \in A; \\ \forall \varepsilon > 0, \exists a_\varepsilon \in A : M - \varepsilon < a_\varepsilon. \end{cases}$

Definition 2. *A non-void real number set B is called lower bounded if there exists $\beta \in R$ so that $\beta \leq b$ for any $b \in B$.*

It can be easily demonstrated that the Axiom of the Upper Boundedness is equivalent to the next **Axiom of the Lower Boundedness:**

For any non-empty lower bounded real numbers set D there exists the highest minimizer in R called the inferior margin of the set D and it is marked with inf (D).

If inf$(D) \in D$, then this property is indicated by inf(D) = min(D).

The following conditions are equivalent:

(i)   and $m = \inf(D)$;

(ii) $\begin{cases} m \leq d, \forall d \in D_1 \\ \forall m_1 \in R \text{ cu } m_1 \leq d, \forall d \in D \Rightarrow m_1 \leq m; \end{cases}$

(iii) $\begin{cases} m \leq d, \forall d \in D, \\ \forall \varepsilon > 0, \exists d_\varepsilon \in D : d_\varepsilon < m + \varepsilon \end{cases}$

The axiom of completeness Cantor-Dedekind is, as well, equivalent to the condition of completeness, understood in the Cauchy manner, of the real number set: *any Cauchy (fundamental) sequence (x_n) of real numbers*

$(\forall \varepsilon > 0, \exists n_\varepsilon \in N : |x_{n+p} - x_n| < \varepsilon, \forall n > n_\varepsilon, \forall p \in N)$ *is convergent to a real*

number.

Theorem 1. *The natural numbers set is not upper bounded.*

Proof.   If N would be upper bounded then, following the Cantor-Dedekind's Axiom of completeness, there exists x_0=sup$(N) \in R$ and for any natural numbers k and n we have:

$$n + k \leq x_0 \tag{1}$$

from where, if we determine that $k \neq 0$, we obtain:

$$n \leq x_0 - k, \ \forall n \in N \tag{2}$$

which shows that $x_0 - k < x_0$ is also an upper bound for N, thus contradicting the fact that x_0 is the lowest upper bound.

Consequence 1.1. *For any real number $x_0 > 0$ there is $m,n \in N^*$ so that*

$$\frac{m}{n} < x_0 \tag{3}$$

Proof.

If the statement wouldn't be true, then there would be

$$x_0 > 0 \text{ with}$$

$$\frac{m}{n} \geq x_0, \ \forall m,n \in N^*. \tag{4}$$

Settling m in (4) and leaving n variable, it results that

$$n \leq \frac{m}{x_0}, \ \forall n \in N^* \tag{5}$$

which contradicts Theorem 1.

Consequence 1.2. (Archimedes' property). *For any real numbers x, $y>0$, there exists $n \in N^*$ so that $nx > y$.*

Consequence 1.3. *For any real number x_0 there exists at least one rational number $r < x_0$.*

Consequence 1.4. *For any real number x_0 there exists at least one rational number s so that*

$$s > x_0 \tag{6}$$

Theorem 2. *For any real number x_0 we have:*

$$x_0 = \sup \{r \in Q : r < x_0\} = \sup \{r \in Q : r \leq x_0\} = \tag{7}$$

$$= \inf \{r \in Q : r > x_0\} = \inf \{r \in Q : r \geq x_0\} =$$

$$= \sup \{x \in R \backslash Q : x < x_0\} = \sup \{x \in R \backslash Q : x \leq x_0\} =$$

$$= \inf \{x \in R \backslash Q : x > x_0\} =$$

$$= \inf \{x \in R \backslash Q : x \geq x_0\}.$$

Proof.

We show that $x_0 = \sup\{r \in Q : r < x_0\}$, the other equalities having similar ways of proofs.

The set $\{r \in Q: r < x_0\}$ is non-empty according to the Consequence 1.3. and, on the basis of Cantor Dedekind's Axiom of completeness, there exists $x' = \sup\{r \in Q : r < x_0\} \in R$.

It is obvious that $x' \leq x_0$ and if $x' < x_0$ then, according to Consequence 1.1 there exists at least a rational number t so that:

$$0 < t < x_0 - x'. \tag{8}$$

On the other hand, because x' is the lowest upper bound of the set $\{r \in Q : r < x_0\}$, there exists $r_0 \in Q$ with $r_0 < x_0$ so that

$$x' - t < r_0. \tag{9}$$

The relation (8) together with (9) leads to

$$x' < r_0 + t < x' + (x_0 - x') = x_0 \tag{10}$$

which shows that $r_0 + t \in \{r \in Q : r < x_0\}$ and contradicts the definition of x'.

Consequence 2.1. *For any two real numbers x_1, x_2 with $x_1 < x_2$ there exists at least one rational number r (irrational s) so that*

$$x_1 < r < x_2 \text{ and } x_1 < s < x_2 \tag{11}$$

Proof.

Indeed, because by virtue of Theorem 2,

$$x_2 = \sup\{r \in Q : r < x_2\} = \sup\{x \in R \backslash Q : x < x_2\}$$ there exists at least a rational number r and at least a irrational one s so that

$$x_1 = x_2 - (x_2 - x_1) < r < x_2 \text{ respectively} \tag{12}$$

$$x_1 = x_2 - (x_2 - x_1) < s < x_2 .$$

Consequence 2.2. *Between any two distinct real numbers there exists an infinity of rational (irrational) numbers and for any real number x_0 there exists at least one increasing sequence and at least a decreasing one made up of only rational (irrational) numbers, both being convergent to x_0.*

Theorem 2 is applied in order to demonstrate this consequence, considering intervals of the type $(x_0 - \frac{1}{n}, x_0)$, $(x_0, x_0 + \frac{1}{n})$ $(n \in N^*)$ and taking into account the elementary property that states that if a real numbers sequence (a_n) is convergent to a real number a and $a_n \geq a$, $\forall n \in N$ ($a_n \leq a$, $\forall n \in N$, respectively), then the order of the elements of the sequence can be changed so that a decreasing (increasing, respectively) sequence convergent to a is obtained.

By the term *"the neighborhood"* of an arbitrary real number x_0 it is understood every set $V \subseteq R$ for which there exists $c, d \in R$, with $c < d$ so that $x_0 \in (c,d) \subseteq V$.

Definition 3. *A non-empty set $D \subseteq R$ is called open if it is for of any proper point, namely*

$$(\forall x \in D,\ \exists (a,b) \subset R\ like\ x \in (a,b) \subseteq D).$$

The family of sets

$$\tau_0 = \{ D \subseteq R : D \text{ is neighbourhood for any of her point} \} \cup \{ \varnothing \}$$

is the *customary topology* on R and can be easily noticed that

τ_0 is closed for the finite intersection and arbitrary reunion.

Actually, *a non-empty real numbers set that doesn't coincide with R is open if and only if it can represented as an at most countable reunion of open non-empty intervals of the type (a,b) or $(-\infty, a)$ or $(b, +\infty)$ with $a, b \in R$, mutually disjoint, the representation being unique up to the order of the terms of the reunion* [4], p.55.

Definition 4. *A set $A \subseteq R$ is called closed if $R\backslash A$ is open.*

Definition 5. *Any point $x_0 \in \bar{R}$ is called adherent point for a non-empty set $A \subset R$ if for any V neighborhood of x_0, we have $V \cap A \neq \varnothing$.*

The set of all the adherent points for a set $A \subseteq R$ is marked with $\bar{A}$.

Theorem 3. *A set $A \subseteq R$ is closed if and only if $A = \bar{A}$.*

Proof.

Let $A \subseteq R$ be a closed set. We mention that for $A=\varnothing$ the property is obvious, and the inclusion $A \subseteq \overline{A}$ takes place every time, thus the only thing left to be demonstrated is $\overline{A} \subseteq A$. Let us consider an arbitrary element $x_0 \in \overline{A}$ and let us absurdly suppose that $x_0 \notin A$. Thus $x_0 \in R \backslash A$ and because $R \backslash A$ is open then there exists a neighborhood V for x_0 so that $V \subseteq R \backslash A$, that is $V \cap A = \varnothing$, which contradicts the membership $x_0 \in \overline{A}$. Thus, $\overline{A} \subseteq A$ and therefore, $A = \overline{A}$.

Reciprocally, suppose that $A = \overline{A}$ and let $x \in R \backslash A$ be arbitrary. Thus $x \notin \overline{A}$, so there exists at least one neighborhood V of x with $V \cap A = \varnothing$, that is, $V \subseteq R \backslash A$.

Therefore, $R \backslash A$ is the neighborhood of x and because x was chosen randomly, it follows that $R \backslash A$ is open.

Theorem 4. *A point $x_0 \in \overline{R}$ is adherent for a non-empty set $A \subseteq R$ if and only if there exists a sequence of points $(a_n) \subseteq A$ with $a_n \to x_0$.*

Proof.

If $x_0 \in R$ is an adherent point for an arbitrary non-empty set $A \subset R$, then by determining the neighborhoods of x_0 being of the type $(x_0 - \dfrac{1}{n}, \ x_0 + \dfrac{1}{n})$, $n \in N^*$, and by minding Definition 5 the existence of a sequence $(a_n) \subseteq A$ with $a_n \to x_0$ is ensured.

Reciprocally, if there exists $(a_n) \subseteq A$ convergent to x_0, then in any neighborhood V of x_0 there is at least one term of the sequence (a_n), thus the intersection $V \cap A$ is non-empty, that is $x_0 \in \overline{A}$. The cases in which $x_0 = -\infty$ ($x_0 = +\infty$) are treated separately, establishing neighborhoods of the type $(-\infty, -n)$ and, respectively $(n, +\infty)$ with $n \in N$.

Generally, if A and B are non-empty sets from an arbitrary topological space, then A is *dense* in B when $B \subseteq \overline{A}$. Consequence 2.2. shows that $\overline{Q} = \overline{R \backslash Q} = R$, that is, the sets Q and $R \backslash Q$ are dense in R. From this point of view the seta Q and $R\backslash Q$ are identical. But, it is well known that the set Q is countable so its Lebesque measure is null, while $R\backslash Q$ has the same cardinal c as the real numbers set, so its Lebesque measure is infinite.

Theorem 5. *A point $x_0 \in R$ is adherent for a non-empty set $A \subset R$ if and only if*

$$d(x_0 ,A) = \inf_{a \in A} | x_0\text{-}a | = 0 \tag{13}$$

Proof.

The definite number provided by (13) is called the *distance from x_0 to the set A.* Consequently this theorem shows that the set of all the adherent points for any non-void set A of real numbers coincides with the real number set situated at a null distance of A.

If $x_0 \in \overline{A}$, then according to the Theorem 4, there exists $(a_n) \subseteq A$ with $a_n \to x_0$, thus $d(x_0,A)=0$.

On the other hand, if (13) occurs, then for any $n \in N^*$ there exists $a_n \in A$ with the property

$$| x_0\text{-}a_n | < \frac{1}{n} \tag{14}$$

and, according to the same theorem, $x_0 \in \overline{A}$.

Theorem 6. *For any two disjoint, closed, non-empty sets A and B from R, there exists two disjoint, open sets G and H so that $A \subset G$ şi $B \subset H$.*

Proof.

For every $a \in A$ we write $d_a=(a, B)$. Due to the fact that $A \cap B = \varnothing$, then $a \notin B = \overline{B}$, and thus according to the previous theorem $d_a > 0$.

In an analogous manner, if for every $b \in B$, we write $d_b = d(b, A)$, then $d_b > 0$. If $x \in R$ and $\varepsilon > 0$ are arbitrary, we mark the open interval symmetrically centered in x of length 2ε with $I(x, \varepsilon)$, that is

$$I(x, \varepsilon) = \{y \in R: | x\text{-}y | < \varepsilon\}.$$

Considering that for any $a \in A$ and any $b \in B$, $I(a, \frac{d_a}{3})$ and $(b, \frac{d_b}{3})$ the result is that the sets $G = \bigcup_{a \in A} I(a, \frac{d_a}{3})$ and $H = \bigcup_{b \in B} I(b, \frac{d_b}{3})$ meet the requirements of the enunciation.

1.5. THE MINIMAL EXTENSION OF R TO $\overline{R}$

In order to achieve a unitary presentation in which any non-empty real numbers set to allow an *infimum*, respectively, a *supremum* one considers the extension of the set R using two *symbols* $-\infty$ attached to any numerical „measure" that becomes "however little" in the sense of the order relation of R and $+\infty$ characterized by the property of every numerical „measure" that "increases however much" in the sense of the same order relation, thus obtaining $\overline{R} = R \cup \{-\infty, +\infty\}$. The geometrical correspondent of this set is called the *extended real straight line*, and the relation of order, the customary algebraic operations and the topology of the real number set are naturally prolonged in $\overline{R}$ as follows:

a) Considering $-\infty < x < +\infty$, $\forall x \in R$ the set $\overline{R}$ is totally ordered.

b) $x + (-\infty) = (-\infty) + x = -\infty$, $\forall x \in \overline{R} \setminus \{+\infty\}$

$x + (+\infty) = (+\infty) + x = +\infty$, $\forall x \in \overline{R} \setminus \{-\infty\}$;

$x \times (\pm\infty) = (\pm\infty) \times x = (\pm\infty)$, $\forall x \in \overline{R}$ with $0 < x$;

$x \times (\pm\infty) = (\pm\infty) \times x = (\mp\infty)$, $\forall x \in \overline{R}$ with $x < 0$;

No sense is attributed to the next situations:

$(-\infty) - (-\infty); \ (+\infty) - (+\infty); \ (+\infty) + (-\infty); \ (-\infty) + (+\infty)$

$0 \times (\pm\infty); \ (\pm\infty) \times 0; \ \frac{(\pm\infty)}{(\pm\infty)}; \ (\pm\infty)^0; \ 1^{(\pm\infty)}; \ 0^0$

because, in every one of these cases, if one considers two numerical sequences which act (from the point of view of the mentioned order relation) exactly as the attached symbols, the result of their composition, in the sense of the indicated operations, is unpredictable.

c) the topology of $\overline{R}$ is induced by the metric

$$\rho: \overline{R} \times \overline{R} \to R_+ \, , \; \rho(x,y) = |\, f(x) - f(y)| \, , \; \forall x,y \in \overline{R}$$

where $f: \overline{R} \to [-1,+1], \; f(x) = \begin{cases} \dfrac{x}{1+|x|}, & x \in R \\ -1, & x = -\infty \\ +1, & x = +\infty \end{cases}$

is *the (limitative) function of René-Louis Baire*

(1874-1932).

Thus the open sets in $\overline{R}$ are only of the following type:

$$G, \; [-\infty, a) \cup G, \; G \cup (b, +\infty], \; G \cup [-\infty, a) \cup (b, +\infty],$$

where G is any open set in R and $a, b \in R$.

1.6. THE IMMEDIATE IMPLICATIONS OF THE AXIOM OF COMPLETENESS CANTOR - DEDEKIND IN CLASSIC MATHEMATICAL ANALYSIS

Following [5], we propose as applications the following properties:

1. Any increasing real numbers sequence has as limit

the superior margin of the set containing its terms:

$$\forall (x_n) \text{ of } R \text{ with } x_n \leq x_{n+1} \, , \; \forall n \in N,$$

$$\exists \lim_{n \to \infty} x_n = \sup\{x_n : n \in N\} = \begin{cases} +\infty, \text{ if } \{x_n : n \in N\} \text{ isn't upper bounded,} \\ s \in R, \text{ if } \{x_n : n \in N\} \text{ is upper bounded.} \end{cases}$$

2. Any decreasing real number set has as limit the inferior margin of the set of its terms:

$$\forall (y_n) \text{ of } R \text{ with } y_n \geq y_{n+1} \, , \; \forall n \in N$$

$$\exists \lim_{n \to \infty} y_n = \inf\{y_n : n \in N\} = \begin{cases} -\infty, & \text{if } \{y_n : n \in N\} \text{ isn't lower bounded,} \\ t \in R, & \text{if } \{y_n : n \in N\} \text{ is lower bounded.} \end{cases}$$

3. Any increasing function $f{:}D{\subseteq}R{\to}R$ has the following lateral limits in any accumulation point x_0 of its definition set D:

$f(x_0-0) = \sup \{f(x){:}\ x \in D \text{ and } x < x_0\}$,

$f(x_0+0) = \inf \{f(x){:}\ x \in D \text{ and } x > x_0\}$.

4. For any decreasing function $f{:}D_1{\subseteq}R{\to}R$ there are lateral limits in any accumulation point y_0 of the definition set D_1 equal to

$f(y_0-0) = \inf \{f(y){:}y \in D_1 \text{ and } y < y_0\}$

$f(y_0+0) = \sup \{f(y){:}y \in D_1 \text{ and } y > y_0\}$, respectively;

Especially,

(i) No matter which would be the increasing function $f : D \subseteq R \to R$ so that $+\infty\,(-\infty)$ is an accumulation point of its definition set D, there exists

$$\lim_{x \to +\infty} f(x) = \sup\{f(x) : x \in D\}$$
$$\left(\lim_{x \to -\infty} f(x) = \inf\{f(x) : x \in D\}\right);$$

respectively

(ii) For any decreasing function $g : E \subseteq R \to R$ so that $+\infty\,(-\infty)$ is an accumulation point of the definition set E, there exists

$$\lim_{x \to +\infty} g(x) = \inf\{g(x) : x \in E\}$$
$$\left(\lim_{x \to -\infty} g(x) = \sup\{g(x) : x \in E\}\right).$$

(iii) Any monotonous real function of real argument has at the most only first rate discontinuities;

(iv) The set of all the discontinuity points of any

monotonous real function of real argument is at the most countable (due to the fact that *the set of the first rate discontinuity points for any real function of real argument is at the most countable - result obtained by the Romanian mathematician Al. Froda (1894 -1973) in 1929)*

5. Let $f:[a,b]\to R$ be a bounded function and $\Delta:(a=x_0<x_1<...<x_n=b)$ an arbitrary division of the interval $[a,b]$. If we write:

$$m_i = \inf\{f(x) : x_i \le x \le x_{i+1}\},\ M_i = \sup\{f(x): x_i \le x \le x_{i+1}\},\ i=\overline{0,n-1},$$

$$s_\Delta(f)=\sum_{i=0}^{n-1} m_i\,(x_{i+1}-x_i),\ S_\Delta(f)=\sum_{i=0}^{n-1} M_i\,(x_{i+1}-x_i)$$

and $\sigma_\Delta(f,\xi) = \displaystyle\sum_{i=0}^{n-1} f(\xi_i)\,(x_{i+1}-x_i)$, where $x_i \le \xi_i \le x_{i+1}, \forall i = \overline{0,n-1}$ is

the Bernhard Riemann's sum attached to the division Δ and to the intermediary points ξ_i, then

$$s_\Delta(f)= \inf\,\{\sigma_D(f,\xi): \xi_i\in[x_i\,,\,x_{i+1}],\ i=\overline{0,n-1}\}\ \text{and}$$

$$S_\Delta(f)= \sup\,\{\sigma_D(f,\xi): \xi_i\in[x_i\,,\,x_{i+1}],\ i=\overline{0,n-1}\}.$$

In addition, the following statements are equivalent:

(i) *the function f is Riemann integrable on* $[a,b]$;

(ii) *for any sequence of divisions* (Δ_n) *of the interval* $[a,b]$ *with* $\|\Delta_n\|\to 0$, *the correspondent Riemann sum sequence, is convergent;*

(iii) *f is almost continuous* (the characterization of Riemann integrability given by Henri Lebesque(1875-1941));

(iv) $\forall \varepsilon > 0, \exists \Delta_\varepsilon$ *- a division of the interval* $[a,b]$ *so that:*

$$S_{\Delta_\varepsilon}(f)-s_{\Delta_\varepsilon}(f)<\varepsilon\,;$$

(v) $\forall \varepsilon > 0, \exists \delta > 0 : S_\Delta(f)-s_\Delta(f)<\varepsilon, \forall \Delta$ *division for* $[a,b]$ *with* $\|\Delta\|<\delta$, *where* $\|\Delta\|=\max\{|x_{i+1}-x_i|:i=\overline{0,n-1}\}$.

Still, at the end of the chapter we must mention some relevant observations regarding the rigorous content and language:

1) the concept of real "domain" is identifiable with every uncommon interval in R;

2) there are no "finite numbers" and "infinite numbers";

3) there are no "cardinal numbers"; the true concept of "number"   preceds, at least historically speaking, the concept of cardinal   expressed through the faculty of sets of "coinciding" due to the same   "power", determined by the bilateral bijections (the customary   equipotence).

CHAPTER 2

Real Functions of Real Arguments

Abstract: This chapter contains a significant sequence of the properties for the real functions of real arguments, also important for the introduction of the fundamental trigonometric functions. Here are included only the properties considered appropriate for defining and describing the essential trigonometric functions. the range of these representative qualities is larger and it is also suggested by the selective bibliography mentioned further on.

Keywords: real function of real argument, graphic, surjective, injective, bijective, inverse, geometric characterization, bounded, monotonous, periodic, centers(axes) of symmetry, Darboux property, continuous, convex(concave).

2.1 INTRODUCTION

Thus, for any real numbers non-empty sets A, B and $f : A \to B$ a function, A is called the *set of definition (domain only if it is non-trivial interval), B is the set where the function gets values (named the co-domain),* and the association $x \in A \to f(x) \in B$, is called the *correspondence law* generated by f.

The set $G(f) = \{(x, f(x)) : x \in A\}$ is the *graphic* of the function f, and what is obtained, when the arranging the elements of $G(f)$ in an arbitrary orthogonal system of Cartesian   coordinates xOy represents the *representation of the graphic* of the function f. It is obvious that every function has a graphic, but there are functions without any graphic representation. Thus, for example, any function $f_{a,b} : R \to \{a, b\}$,

defined by $f_{a,b}(x) = \begin{cases} a, x \in Q \\ b, x \in R \setminus Q \end{cases}, \forall a, b \in R, a \neq b.$

If g is a real function of real argument, that can be represented graphically, then the graphic representation of the function $f = |g|$ coincides with that corresponding to the function g for $g(x) \geq 0$ and it is obtained from this by the symmetry with respect to the axis Ox in all the points x for which $g(x) \leq 0$. For every function $f : A \to B$ and any non-empty sets $E_1 \subseteq A \subseteq E_2$, the function $f_{/E_1} : E_1 \to B$ with the property: $f_{/E_1}(x) = f(x),\ \forall x \in E_1$ is called the *restriction* of the function f to the set E_1 and any of the functions $\tilde{f} : E_2 \to R$ having the restriction $\tilde{f}_{/A} = f$ is called the *extension* of the function f to the set E_2. Under the mentioned circumstances, $f(E_1) = \{f(x) : x \in E_1\}$ is called *the image* of the set E_1 through   the   function f and   for   every   non-empty   set $B_1 \subseteq B, f^{-1}(B_1) = \{x \in A : f(x) \in B\}$ is called the *counter-image* of the set B_1 through the function f.

Thus, by exemplifying, the restriction of the "whole part function" $[\ ] : R \to Z$ defined by

$$[x] = \begin{cases} x, x \in Z \\ \text{the highest integer number strictly lower than } x, x \in R \setminus Z \end{cases}$$

and characterised by the simultaneous accomplishment of the following requirements:

$$\begin{cases} [x] \in Z, \\ [x] \leq x, \\ [x] + 1 > x, \forall x \in R \end{cases}$$

in the integer numbers set coincides with the identical function $1_Z : Z \to Z$, $1_Z(u) = u, \forall u \in Z$, being in the same time an extension to the real number set of this function, because $[R] = Z$, and $[Z]^{-1} = R$.

2.2. SURJECTIVE, INJECTIVE, BIJECTIVE FUNCTIONS

A function $f : A \to B$ is *surjective* when $f(A) = B$, that is for every $y \in B$ the equation $f(x) = y$ has at least one solution in set A. If for any $x_1, x_2 \in A$ with $x_1 \neq x_2$, $f(x_1) \neq f(x_2)$, this means that any coincidence $f(a) = f(b)\,(a, b \in A)$ implies only $a = b$, then f is called *injective*.

Graphically speaking, a function $f : A \to B$ is surjective if and only if *any parallel to the Ox axis through the set B represented on the Oy axis crosses the representation of the graphic at least in one point* and it is injective when and only when *any parallel to the Ox axis crosses the representation of the graphic in one point at the most.* Thus, the bijective quality of a random function $f : A \to B$, when it exists, *is characterized by the property of any parallel to the Ox axis by the elements of the set B represented on the Oy axis, of crossing the representation of the graphic of the function f in exactly one point.* Generally speaking, there is no relation between surjectivity and injectivity, as it is shown by the following examples.

1. The function $f_1 : R \to R_+ = [0, +\infty)$, $f_1(x) = x^2$ is surjective, but isn't injective because $f_1(-2) = f_1(2) = 4$ and $-2 \neq 2$ (see Figure **1.**)

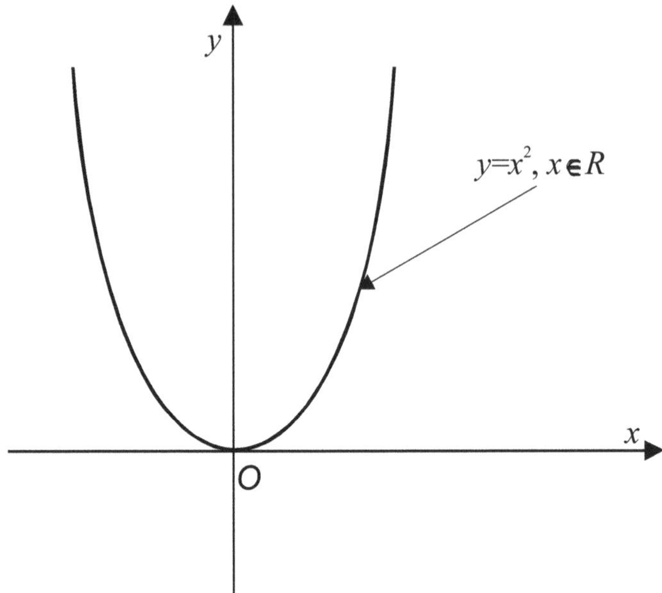

Figure 1: Surjectivity without injectivity.

2. The function $f_2:(-\infty,0]\to R$, $f_2(x)=x^2$ is injective, but the equation $y=f_2(x)$ doesn't have solutions for $y\in(-\infty,0)$, thus f_2 isn't a surjective function.

3. The function $f_3:R\to R$, $f_3(x)=\begin{cases}\dfrac{1}{x}, & x\in R\setminus\{-1,0,1\}\\[2mm] 2, & x\in\{-1,0,1\}\end{cases}$

(see Figure 2.)

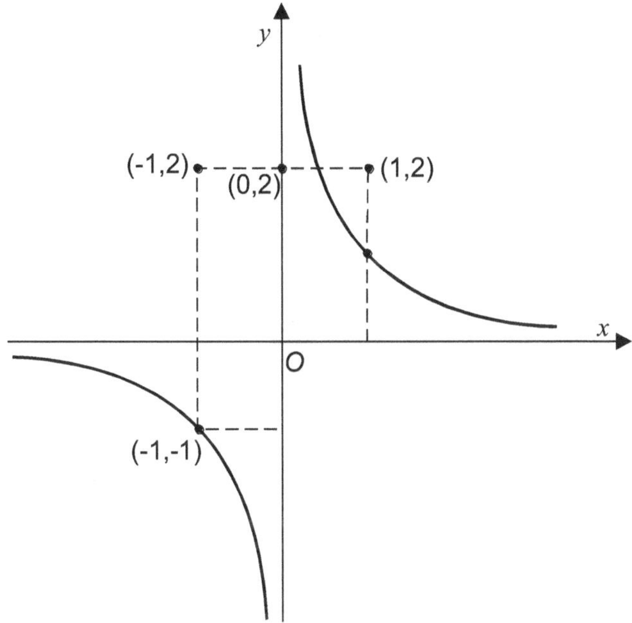

Figure 2: Not surjective or injective function.

The function f_3 is neither surjective, nor injective. The bijectivity of a function $\forall x \in A \xrightarrow{f} y = f(x) \in B$ characterizes the existence of the inversion $\forall y = f(x) \in B \xrightarrow{f^{-1}} x = f^{-1}(y) \in A$, and the graphic representation of the inversion $f^{-1} : B \to A$ is obtained through the graphic representation of the function $f : A \to B$ (anytime it exists) through symmetry to the first bisecting line.

For example, the inverse function corresponding to $f_4 : [0, +\infty) \to [0, +\infty)$, $f_4(x) = x^2$ is $f_4^{-1} : [0, +\infty) \to [0, +\infty)$, $f_4^{-1}(x) = \sqrt{x}$ and can be represented graphically in the following manner (see Figure 3.)

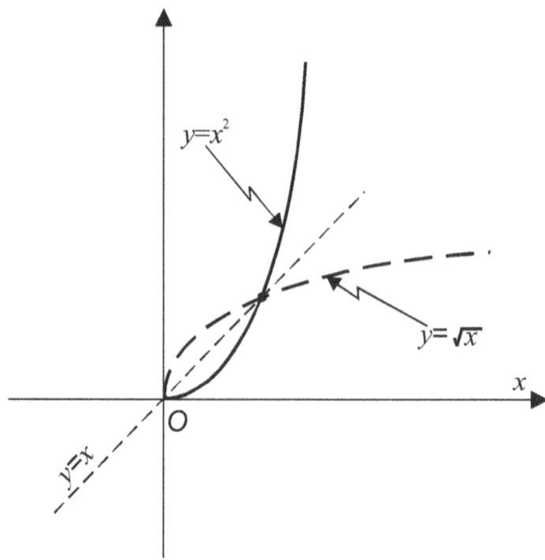

Figure 3: Geometrical representation of an inverse function.

2.3. BOUNDED (UNBOUNDED) FUNCTIONS

A real function of real argument $f : A \rightarrow B$ is called *bounded* if it meets
the one of the following equivalent requirements:

      (i)   the set $f(A) = \{f(x) : x \in A\}$ is bounded;

      (ii) $\exists \alpha, \beta \in R : \alpha \leq f(x) \leq \beta, \ \forall x \in A$;

      (iii) $\exists T \geq 0 : |f(x)| \leq T, \ \forall x \in A$,

and *unbounded*, respectively, on the set A if it does not satisfy any of
this requirements.

Thus, the function $f_5 : R \rightarrow [0, +\infty)$, $f_5(x) = \dfrac{|x|}{1+|x|}$ is bounded because

$0 \leq \dfrac{|x|}{1+|x|} < 1, \ \forall x \in R$, and $f_6 : R^* \rightarrow R$, $f_6(x) = \dfrac{1}{x}$ is unbounded, because,

if we were to admit that f_6 meets the requirement (iii), then there would

exist $T > 0$ so that $1 \leq T|x|, \ \forall x \in R^*$, which for $x = \dfrac{1}{2T}$ leads to a

contradiction.

2.4. STRICTLY MONOTONOUS (MONOTONOUS) FUNCTION

If $f : A \subseteq R \to B \subseteq R$, is an arbitrary function, then

    a) f is *strictly increasing* on $A \Leftrightarrow f(x_1) < f(x_2)$, $\forall\ x_1, x_2 \in A$ with $x_1 < x_2$.

Example: $x \to f_{s,t}(x) = tx + s$, $t > 0$, $s \in R$

    b) f is *increasing* on $A \Leftrightarrow f(x_1) \leq f(x_2)$, $\forall\ x_1, x_2 \in A$ with $x_1 < x_2$.

Example: Any constant function. It is obvious that every strictly increasing function is increasing on the set of definition, but there are increasing functions that aren't strictly increasing (for example, any constant function).

    c) f is *strictly decreasing* on $A \Leftrightarrow f(x_1) > f(x_2)$, $\forall\ x_1, x_2 \in A$ with $x_1 < x_2$.

Example: $x \to g_{c,d}(x) = cx + d$, $c < 0$, $d \in R$.

    d) f is *decreasing* on $A \Leftrightarrow f(x_1) \geq f(x_2)$, $\forall\ x_1, x_2 \in A$ with $x_1 < x_2$.

Any strictly decreasing function is decreasing, but there are decreasing functions that aren't strictly decreasing (for example, any function $h = -u$ with u an arbitrary non-injective increasing function).

Any function f that meets the requirements a) or c) is called *strictly monotonous*, and if it satisfies b) or d) then it is called *monotonous* on the set A. The monotony of a function on a compact interval $I = [a,b] \subset R$ warrants the boundedness of the function, but, in general, the boundedness property does not necessarily imply the monotony; thus the function meets this requirement.

$$f_7 : [-1,1] \to [0,1] \text{ defined by } f_7(x) = \begin{cases} 1+x, & x \in [-1,0] \\ 1-x, & x \in [0,1] \end{cases} \quad \text{(see Figure 4.)}$$

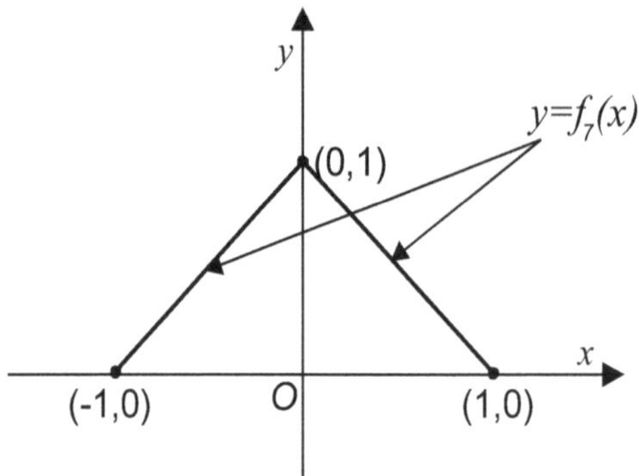

Figure 4: Boundedness without monotonicity.

Any strictly monotonous function is, obviously, injective, but there are injective functions that are not necessarily strictly monotonous.

This is the case for the function

$f_8 : [-1, +\infty) \to (-\infty, 1]$ given by

$$f_8(x) = \begin{cases} x+1, & x \in [-1, 0] \\ -x, & x > 0 \end{cases} \text{(see Figure 5.)}$$

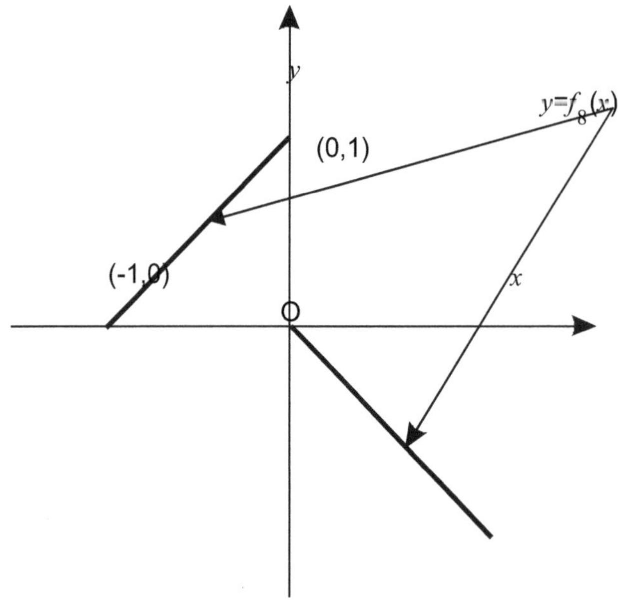

Figure 5: Injective function without monotonicity.

2.5. PERIODIC FUNCTIONS

A function $f : A \subseteq R \to B \subseteq R$ is called *periodic* if there exists $\alpha \neq 0$ so that $f(x + \alpha) = f(x)$, $\forall x, x + \alpha \in A$.

The lowest strict positive real number α with the above mentioned property (if it exists), is called *the principal period* of the function, and the graphic representation of the function, when it is possible, is obtained from the graphic (restriction) representation of the f on the interval $[0, |\alpha|]$ through succesive translations of vector with the lenght $|\alpha|$.

If α is a period of the function f, then by using the adequate extensions for f, it can be inductively observed that $f(x + n\alpha) = f(x - n\alpha) = f(x)$, $\forall x \in A$, $n \in N$. Not every periodic function has a principal period. In order to support this we, consider that $t_1, t_2 \in R$, $t_1 \neq t_2$ and the function $f_{t_1, t_2} : R \to \{t_1, t_2\}$ having the

correspondence law $f_{t_1, t_2}(x) = \begin{cases} t_1, & x \in Q \\ t_2, & x \in R \setminus Q \end{cases}$

Every non-null rational number is period for

f_{t_1, t_2}, but $\inf(Q) = -\infty$

Clearly, any periodic, monotonous, real function on R is , obviously, a constant function and conversely.

2.6. CENTERS AND AXES OF SYMMETRY

In order to approximate the graphic representation of a real function of real argument, every time it exists, sometimes noticing the centers, respectively, the axes of symmetry is also important.

(x_0, y_0) is a *symmetry center* for the graphic $G(f) = \{(x, f(x)) : x \in A\}$ of a function $f : A \subseteq R \to B \subseteq R$ if $f(x_0 - t) + f(x_0 + t) = 2y_0$, $\forall t \in R$ with

$x_0 - t, x_0 + t \in A$, which, for the graphic representation, when it exists, means that the symmetric of any point belonging the graphic representation in relation to (x_0, y_0) preserves this belonging. An interesting particular case is represented by the odd functions $f : A \subseteq R \to B \subseteq R$ characterised by the relation $f(-x) = -f(x)$,

$\forall x \in A$ with $x \in A$, for which $(0,0)$ is symmetry center.

For example, $f_9 : R \to R$, $f_9(x) = x^{2n+1}$, $(n \in N)$ (see Figure 6.)

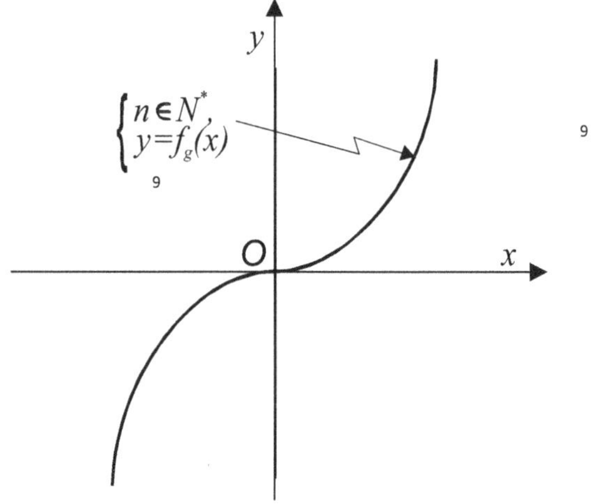

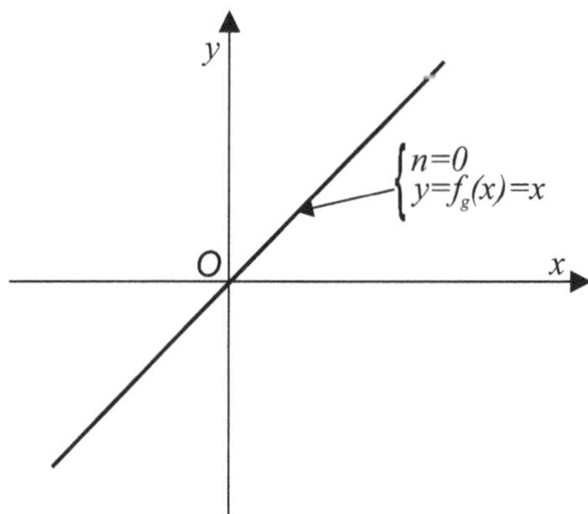

Figure 6: Ilustrated example 1.

Under the same circumstances, the graphic of a function $f : A \subseteq R \to B \subseteq R$ allows as *symmetry axis* the equation straight line $x = t$ if $f(t-s) = f(t+s)$, $\forall t-s,\ t+s \in A$.

Geometrically speaking, this is the replacement of the precedent symmetry in relation to the (x_0, y_0) point with the symmetry to the straight $x = t$. Thus, the graphic of any even function $f : A \subseteq R \to B \subseteq R$ characterized by the property $f(-x) = f(x)$, $\forall x, -x \in A$ allows as symmetry axis the equation straight line $x = 0$, that is the O_y axis, as it can be immediately observed also for any function $f_{10} : R \to [0, +\infty)$, $f_{10}(x) = x^{2n}$, $n \in N$ (see Figure 7.)

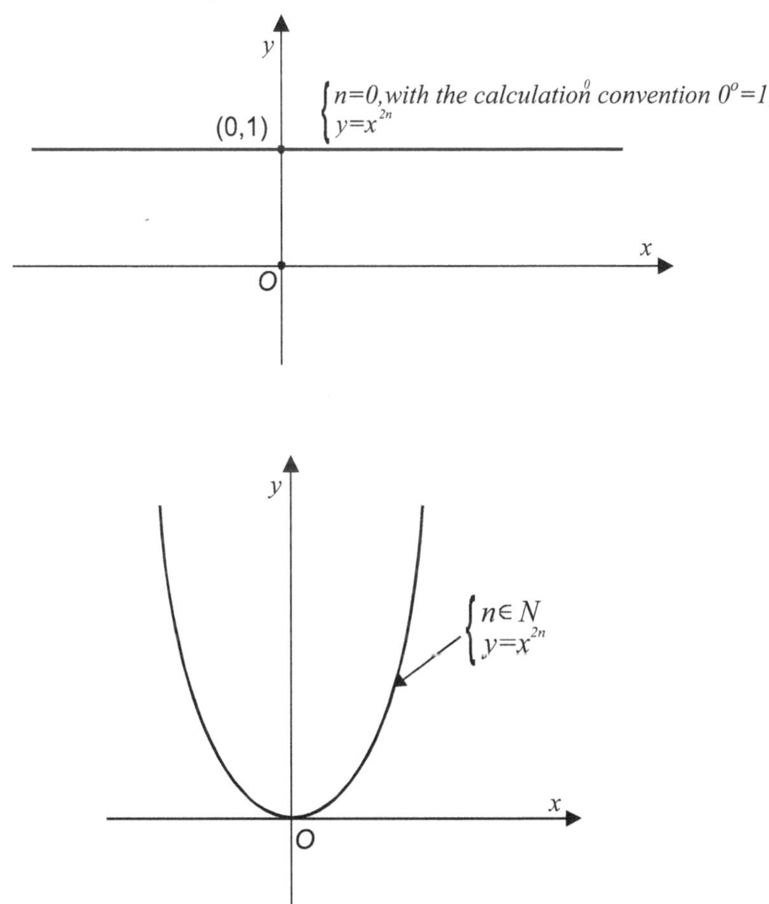

Figure 7: Ilustrated example 2.

respectively   for   the   module   function $f_{11}: R \to [0, +\infty)$, $f_{11}(x) = |x|$ graphically represented in the following manner (see Figure 8.)

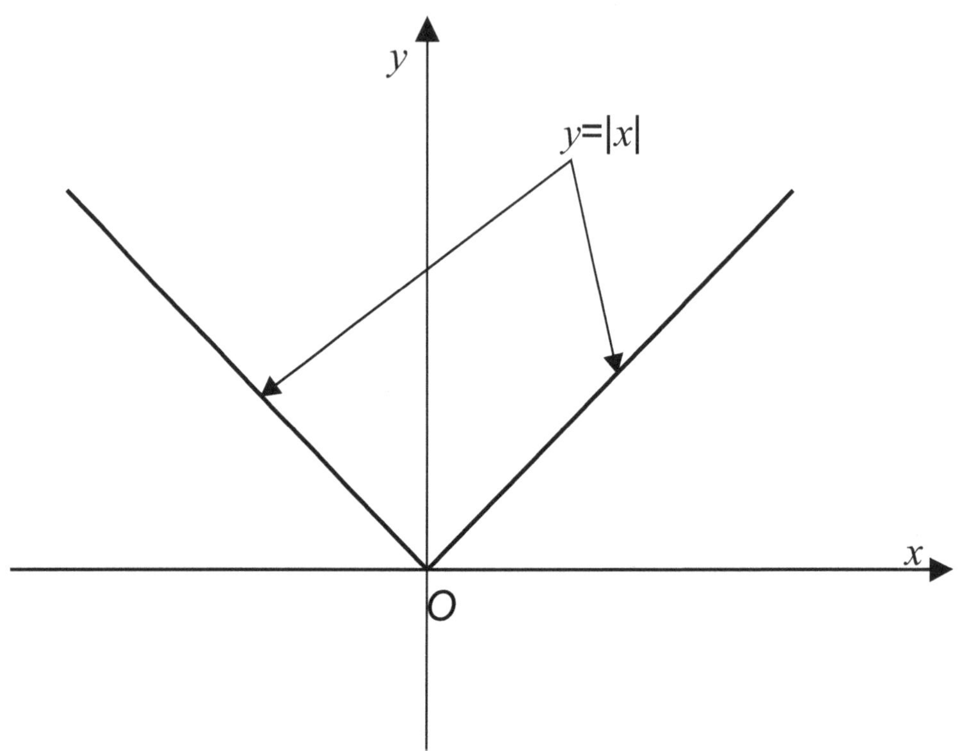

Figure 8: Ilustrated example 3.

2.7. FUNCTIONS WITH THE PROPERTY OF JEAN GASTON DARBOUX   (1842-1917)

A   function $f: A \subseteq R \to B \subseteq R$ has   the   *Darboux property* (the property of intermediate value), if no matter what the $a, b \in A$, $a < b$ and no matter what the real number λ between $f(a)$ and $f(b)$, there is at least   one   point c_{λ} between a and b so that $f(c_{\lambda}) = \lambda$, that   is $[f(a), f(b)] \subseteq f(A \cap [a,b])$ or $[f(b), f(a)] \subseteq f(A \cap [a,b])$ which   is equivalent with the property of $f(I)$ being an interval for any interval $I \subseteq A$, but not necessary of the same type.

Other examples and related properties

1. Any function $f_{c,d} : R \to R$, $f_{c,d}(x) = cx + d$, $c, d \in R$

being arbitrary determined, satisfies the

Darboux's property.

2. $f_{12} : [0,6] \to R$, $f_{12}(x) = \begin{cases} -x+3, & x \in [0,3] \\ x, & x \in (3,6] \end{cases}$ doesn't

have Darboux's property because

$1 = f(2) < 2 < f(5) = 5$, and the equation $f(x) = 2$

allows the solution $x = 1 \notin (2,5)$ respectively $x = 2 \notin (3,6]$.

3. $f_{13} : R \to R$, $f_{13}(x) = \begin{cases} x, & x \in Q \\ x^2, & x \in R \setminus Q \end{cases}$ doesn't

satisfy the Darboux property

because $1 = f(1) < 4 < f(\sqrt{5}) = 5$, and the equation

$f(x) = 4$ doesn't have solutions in the interval $[1, \sqrt{5}]$.

4. Any real function of real argument, defined on an uncommon interval, can be represented as the sum of two functions with the Darboux property (the theorem of Waclaw Sierpinski(1882-1969).

5. Any injective function with Darboux property on an arbitrary non-trivial interval *I* is strictly monotonous.

Indeed, if we were to admit that there exists an injective function $f : I \subseteq R \to R$ with the Darboux property, which isn't strict monotonous, then there would exist $x_1, x_2, x_3 \in I$ with $x_1 < x_2 < x_3$ and $f(x_2)$ is not placed between $f(x_1)$ and $f(x_3)$, that is supposing that $f(x_1) < f(x_3)$ we would obtain

$f(x_2) < f(x_1) < f(x_3)$ or $f(x_1) < f(x_3) < f(x_2)$ which when combined with the Darboux property for the function f implies the existence of the points c_1 between x_2 and x_3 or c_2 between x_1 and x_2 so that $f(c_1)=f(x_1)$ or $f(c_2)=f(x_3)$, both being contrary to the injectivity of the function f.

6. According to the definition of the function that have the property of the intermediary value, any function $f:[a,b]\to R$ $(a,b\in R,\ a<b)$ having the Darboux property, and $f(a)\cdot f(b)<0$ is anulled in at least a point $c\in(a,b)$. Hence, if $f:I\subseteq R\to R$ has the Darboux property on the interval I and $f(x)\neq 0,\ \forall x\in I$, then $f(x)<0,\ \forall x\in I$ or $f(x)>0,\ \forall x\in I$.

7. Let $a,b\in R,\ a<b$ and $f:[a,b]\to[a,b]$ an application with the property that the function $g:[a,b]\to R,\ g(x)=f(x)-x$ has the Darboux property. In this case the simultaneous inclusions are not possible: $f\left(Q\cap[a,b]\right)\subseteq(R\setminus Q)$

and $f\left((R\setminus Q)\cap[a,b]\right)\subseteq Q$.

Indeed, supposing the contrary,
$g(a)=f(a)-a\geq 0,\ g(b)=f(b)-b\leq 0$. From the Darboux property owned by the function g on the interval $[a,b]$ there exists $x_0\in[a,b]$ so that $g(x_0)=0\Leftrightarrow f(x_0)=x_0$, a contradiction because if $x_0\in Q\cap[a,b],\ f(x_0)\in R\setminus Q$ and $x_0\in(R\setminus Q)\cap[a,b]$ implies $f(x_0)\in Q$.

8. There are functions $f:D\subseteq R\to R$ with the Darboux property so that $f(D\cap Q)\subseteq R\setminus Q$ and

$$f\left[D\cap(R\setminus Q)\right]\subseteq Q\ ?$$

This original mentioned issue has the following solution : If the definition set D would include at least non-trivial interval I, then $f(I) = J$ would be, according to Darboux property, valid for f, and also an uncommon interval, not necessarily having the same structure as the interval I. But $J = f(I) = f\left[I \cap (Q \cup R \setminus Q)\right] = f(I \cap Q) \cup f\left[I \cap (R \setminus Q)\right]$ is an impossible equality because J has the same cardinal as any interval in R which isn't reduced to a point, and $f(I)$ is, in this situation, a countable set at the most. Thus, anytime D includes at least on uncommon interval, the answer is negative. Examining the other cases is now simple.

2.8. CONTINUOUS FUNCTIONS

An important particular case of functions with Darboux property is represented by the continuous function.

Let $D \subseteq R$ be a non-empty set, and $x_0 \in D$ arbitrary, $\mathcal{V}(x_0) = \{V \subseteq R : \exists (a,b) \text{ with } x_0 \in (a,b) \subseteq V\}$ the set of all the *neighbourhoods* of the element x_0.

A function $f : D \to R$ is called *continuous* in x_0 if $\exists \alpha, \beta \in R$, $\alpha < \beta$ with $(\alpha, \beta) \cap D = \{x_0\}$ or for every neighbourhood U of the number $f(x_0)$, there exists $V \in \mathcal{V}(x_0)$ so that $f(V \cap D) \subseteq U$.

When this local property of the function f is generalized in every point in D, f is called *continuous* in D, fact characterized for $D = R$ by every one of the following equivalent requirements, taking into account the statements presented in Chapter I.

  a) the counter-image of any open set is an open set;

  b) the counter-image of any closed set is a closed set;

  c) $f(\bar{M}) \subseteq \overline{f(M)}$, $\forall M \subseteq R$ with $M \neq \phi$ where

$\overline{M(f(M))}$ *represents the closure of the set* $M(f(M))$. *Any function which isn't continuous in* $x_0 \in D$ *(on* D*) is called discontinuous in* x_0 *respectively on* D.

Both the problem of continuity and that of discontinuity has sense only in the points where the function is defined.

Immediate examples and properties

1. $f_{14} : R \to [0,\infty)$, $f_{14}(x) = |x|$, $\forall x \in R$ continuous

   on R;

2. $f_{15} : R \to R$, $f_{15}(x) = \begin{cases} x, & x \in Q \\ -x, & x \in R \setminus Q \end{cases}$ continuous

   only on origin

3. Any function $f : [a,b] \to R$ continous on an arbitrary compact interval $[a,b]$ is bounded and reaches its limits, that means that there exist $x_1, x_2 \in [a,b]$ so that

   $f(x_1) = \min\{f(x) : a \le x \le b\}$, and

   $f(x_2) = \max\{f(x) : a \le x \le b\}$.

4. The continuity of any real function of real argument f on any uncommon interval I in R ensures the validity of Darboux property for f on I. Still there are totally discontinuos functions having the Darboux property, indicated by Darboux himself.

5. The inversion of any continuos and injective function on an arbirtrary, non-trivial interval in R is continous and strict monotonous.

2.9. CONVEX (CONCAVE) FUNCTIONS

Let be $I \subseteq R$ be a non-trivial interval. A function $f : I \to R$ is called *convex* on I if it meets one of the following equivalent requirements:

(i) $f(\alpha x_1 + \beta x_2) \le \alpha f(x_1) + \beta f(x_2),$

$\forall x_1, x_2 \in I,\ \alpha, \beta \ge 0$ cu $\alpha + \beta = 1$;

(ii) $\forall n \in N^*,\ n \ge 2,\ x_1, x_2, ..., x_n \in I,\ \alpha_1, \alpha_2, ..., \alpha_n \ge 0$

and

$$\sum_{i=1}^{n} \alpha_i = 1 \Rightarrow f(\sum_{i=1}^{n} \alpha_i x_i) \le \sum_{i=1}^{n} \alpha_i f(x_i);$$

(iii) $\forall x_1, x_2, x_3 \in I$ and

$$x_1 < x_2 < x_3 \Rightarrow \frac{f(x_2) - f(x_1)}{x_2 - x_1} \le \frac{f(x_3) - f(x_2)}{x_3 - x_2};$$

(iv) $\forall x_1, x_2, x_3 \in I$ with $x_1 < x_2 < x_3 \Rightarrow$

$$\begin{vmatrix} 1 & 1 & 1 \\ x_1 & x_2 & x_3 \\ f(x_1) & f(x_2) & f(x_3) \end{vmatrix} \ge 0$$

If the last inequality mentioned at every of the points (i) – (iv) is strict, than the function f is called *strictly convex* on the interval I.

Every time this inequality changes its sense (respectively, it becomes strict) the function at which it refers is called *concave (strictly concave)*.

Generally, a function is convex (strict convex) on I if and only if the function $-f$ is concave (strict concave) on the same interval.

Examples

1. $f_{16} : R \to R_{+1}$

$$f_{16}(x) = \begin{cases} x^2, & x \in R \setminus [-1,1] \\ 1, & x \in [-1,1] \end{cases}$$ is covex on R (see Figure 9.)

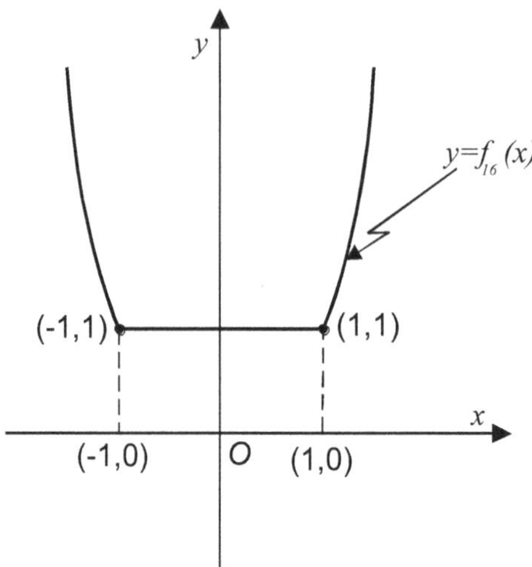

Figure 9: Convex function.

2. $f_{17} : R \to R_{1}$, $f_{17}(x) = x^2$ is strictly convex on R

(see Figure 10.)

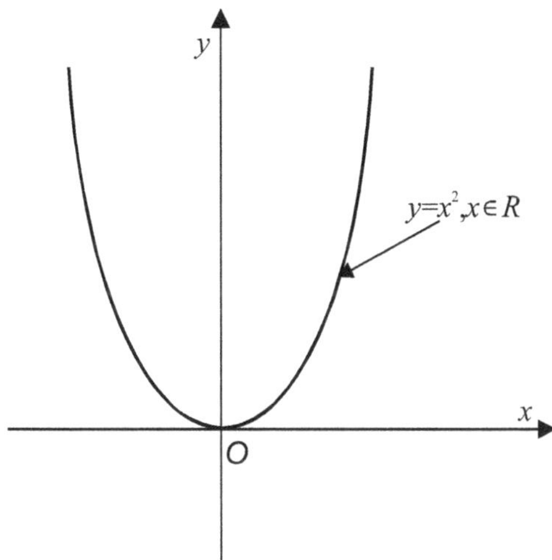

Figure 10: Strictly convex function.

3.　$f_{18} : R \to R_1,$　$f_{18}(x) = x^3$　is　strictly concave on $(-\infty, 0]$ and strictly convex on $[0, +\infty)$ (see Figure 11.)

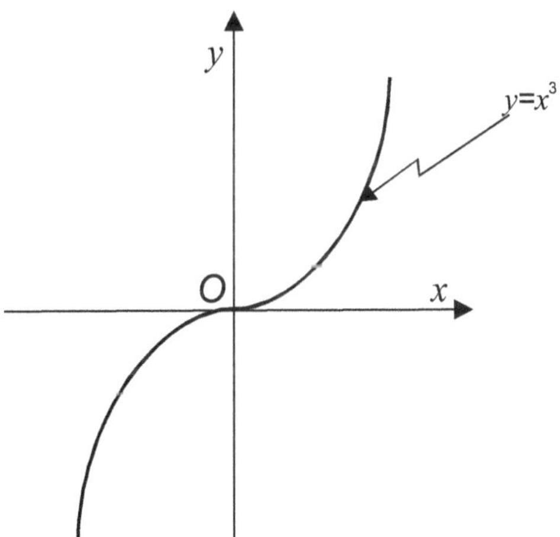

Figure 11: Strictly concave and strictly convex function.

4. $f_{19} = -f_{16}$ is a concave function on the real number Set (see Figure 12.)

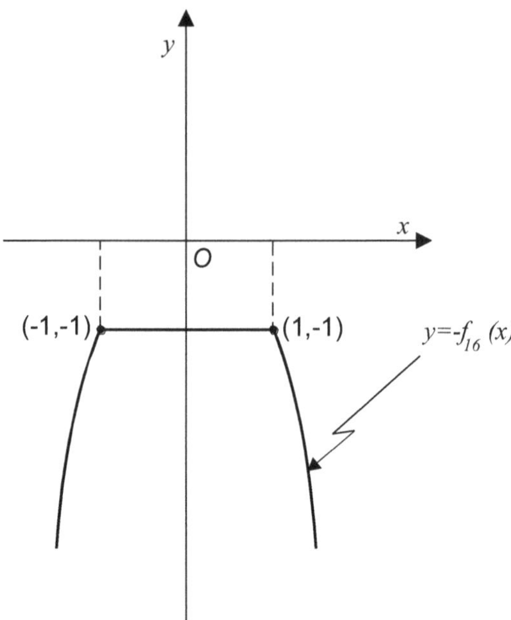

Figure 12: Concave function.

All the following functions specific to the Trigonometry will be examined and described under the previous scenario.

CHAPTER 3

Vectors in R3

Abstract: In the beginning we will present the first complete axiomatic system for Euclidean geometry elaborated in 1899 by David Hilbert as a sinthesis of Lectures on the Foundations of Geometry, 1891-1902.

Keywords: Linear(vector) space, foundation of the finite dimensional geometry in the three dimensional space, Euclidean's moment, Hillbert's system of axioms, scalar product, generated norm, properties, examples.

3.1. INTRODUCTION

Hilbert requires the existence of three groups of „objects" called points, straight and planes without any previous descriptions of the objects of geometric study, and by *geometric space* he means a set of elements subjected to some relations that respect the requirements imposed by the axioms.

The objects of the first group called points are determined and are marked with: A, B, C, ...

The objects of the second group are called straights and are marked with $a, b, c, \dots$ and those of the third group called *planes* are marked with $\alpha, \beta, \gamma, \dots$ __

The points and the straights form the elements of the plane geometry, and the points, straights, and planes are the elements of the solid geometry.

3.2. AXIOMS OF INCIDENCE

1. *For any two different points there exists at least one straight line to which they belong.*

2. *For any two distinct points there is at the most a straight line to which they belong.*

3. *For any straight line, there are at least two distinct points which belong to it. In any plane there are at least three points that don't simultaneously belong to the same arbitrary straight line included in the plane.*

4. *For every mutually different points A, B, C there is at least one plane to which they belong. In any plane there is at least one point.*

5. *For any three distinct points there is at most one plane to which they belong.*

6. *If two distinct points of an arbitrary straight line belong to a plane, then any point of the straight line belongs to the plane.*

7. *If two planes have a common point, then they have at least one more point in common.*

8. *There are at least four points*

   that aren't in the same plane.

Consequences.

1. *Any two different straight lines have at the most one point*

   in common.

2. *Either two planes don't have any point in common or have a common straight line or they coincide.*

3. *Any plane and any straight line, that doesn't belong in that plane, have one point in common and the most.*

4. *Any straight line and any point exterior to the straight line, as well through two concurrent straight lines passes an unique plane.*

5. *Every there distinct points generate an unique plane.*

6.   *In any plane there are at least three points that are not colinear.*

7.   *There exists at least one point exteriorto any plane.*

3.3. AXIOMS OF ORDER

1.   *If the point B is between points A and C then the points A,B,C are colinear and B is between C and A : $A-B-C \Rightarrow A,B,C$ colinear and $C-B-A$.*

2.   *For any two points A and B there is at least one point C so that $A-B-C$.*

3.   *Of any three distinct points of an arbitrary straight line one and only one point is between the other two.*

4.   *(Pash's axiom) If an arbitrary straight line crosses in the interior one of the sides of an uncommon triangle, then it also crosses on the interior any other side of the triangle.*

     (Equivalent expression: If A,B,C are three arbitrary non-colinear points and d a random straight line in the plane (A,B,C) with $d \cap |AB| \neq \varnothing$, then $d \cap |AC| \neq \varnothing$ or $d \cap |BC| \neq \varnothing$)

Consequences.

1.   *Any uncommon segment has an infinity*

of points.

2.   *Of three arbitrary distinct points of a any straight line there is one situated between the other two.*

3.   *If an arbitrary straight line crosses two of the arbitrary segments $|AB|,|BC|,|CA|$, then it can't cross a third segment no matter what would be the non-colinear points A, B, C.*

4.   *On the support straight line of any uncommon segment there is an infinity of points situated in the interior of the segment as well as an infinity of points situated in the exterior.*

5. **The Division Property of the Straight Line.** *Any point of any straight line divides the straight line in two non-empty disjoint subsets called semi-straights, one being closed (which includes the point) and the other open, so that their reunion coincides with the setof the points of the straight line (thus, a partition of the straight is achieved).*

6. **The Division Property of the Plane.** *In any plane any straight divides the set of the plane points in two non-empty disjoint subsets called semi-planes, one closed (which includes the straight line), and the other open, so that their reunion uses up the set of the plane points.*

7. **The Division Property of the Space.** *Any plane divides the set of the points of any customary Euclidean space in two semi-spaces, one closed (which includes the plane), and the other open, so that a partition of the respective space is achieved, that is every one of these sets is non-noid, being disjoint from the other, and by reunion it coincides with the set of all the space points.*

3.4. AXIOMS OF CONGRUENCE

Two arbitrary non-empty sets Λ and Λ' in space are called *congruent*, marked with $A \equiv A'$, if at least one bijection from A to A' that preserves the distance between points, can be indicated.

Any relation of congruence is obviously *reflexive*, *symmetrical* and *transitive*, that is, a genuine relation of equivalence.

1. *On any semi-straight there is an unique segment congruent to an arbitrary uncommon segment and having one extremity in the origin of the semi-straight line.*

2. *Any two segments congruent with a third one are congruent between them:*

$$\forall \left|AB\right|, \left|A'B'\right| \ with \left|AB\right| \equiv \left|CD\right| \ and$$

$$\left|A'B'\right| \equiv \left|CD\right| \Rightarrow \left|AB\right| \equiv \left|A'B'\right|$$

3. *The axiom of segment addition:*

$$\forall (A,B,C),(A',B',C'), A-B-C, A'-B'-C'$$

$$\text{and } |AB| \equiv |A'B'|, |BC| \equiv |B'C'| \Rightarrow |AC| \equiv |A'C'|$$

4. *In any semi-plane there is an unique angle congruent with a given angle, with the vertex in a point belonging to the origin straight line of the semi-plane (by "angle" is understood any couple of two semi-straights with a common origin).*

5. $\forall (A,B,C),(A',B',C')$ *with* $\begin{cases} |AB| \equiv |A'B'| \\ |AC| \equiv |A'C'| \\ \widehat{BAC} \equiv \widehat{B'A'C'} \end{cases} \Rightarrow \begin{cases} |BC| \equiv |B'C'| \\ \widehat{ACB} \equiv \widehat{A'C'B'} \\ \widehat{ABC} \equiv \widehat{A'B'C'} \end{cases}$

For I-III immediate consequences are the

customary cases of triangles congruence.

3.5. THE ARCHIMEDE'S AXIOM

If $|AB|$ *and* $|CD|$ *are two arbitrary segments, then on the straight line* AB *there is a finite set of points* $\{A_1, A_2, A_3,.., A_n\}$ $(n \in N*)$ *so that* $A-A_1-A_2, A_1-A_2-A_3$ *and so on, the segments* $|AA_1|, |A_1A_2|,..., |A_{n-1}A_n|$ *are congruent with the segment* $|CD|$ *and* $A-B-A_n$.

3.6. CANTOR'S AXIOM

Being given an arbitrary straight line, which contains a descending sequence $|A_1B_1|, |A_2B_2|,..., |A_nB_n|,..$ *(that means that every segment is included in the interior of the preceding one, except for the first one), and there is no segment situated in the interior of all the segments of the sequence, then on the straight line there is a unique point included in the interior of all the segments.*

3.7. AXIOM OF PARALLELS

In any exterior point of any straight line there is a unique parallel to that straight line.

Statements equivalent to the Axiom of parallels:

(P1) The sum of the interior angles of any uncommon triangle coincides with two right angles.

(P2) All non-degenerated triangles have the same sum of the interior angles.

By *a Dedekind section* on any straight line d , oriented by a order relation "<" is understood any subsets couple (Σ, λ) of the straight line, with the following properties:

$$\Sigma, \lambda \neq \varnothing$$

$$\Sigma \cap \lambda = \varnothing$$

$$\Sigma \cup \lambda = d$$

$$\forall T \in \Sigma, S \in \lambda \Rightarrow T < S$$

Any Dedekind section on (d,<) for which there exists a point that divides Σ and λ is called *first rate Dedekind section*. The Dedekind sections that (d,<) aren't of first rate are called of *second rate*.

Let's take in consideration the following statements:

(D1 For any Dedekind section on (d,<)

and any segment $|CD|$, $\exists A \in \Sigma$ and $B \in \lambda$ so that $|AB| \equiv |CD|$.

(D2) If a Dedekind section on (d,<) satisfies (D1), then it is of first rate.

(D1) and (D2) are equivalent **to The Dedekind Principle** (Theorem V.33, [6]):

Every Dedekind section on any oriented straight line (d,<) is of first rate.

Furthermore, the axioms 3.4 and 3.5 can be replaced by the Dedekind principle, without affecting Hilbert's system of axioms.

Details regarding the axiomatic construction   indicated by David Hilbert for Euclidean geometry and also extremely useful commentaries will be found in Chapter V from [6].

A non-empty set X is called linear (vectorial) space over a field $(K,+,\cdot)$ if in relation to an internal binary operation $\oplus: X \times X \to X$ is an Abelian group and an external operation $\otimes: K \times X \to X$ satisfies the following properties:

$$1)\ \alpha \otimes (x \oplus y) = \alpha \otimes x \oplus \alpha \otimes y,\ \forall x, y \in X,\ \alpha \in K;$$

$$2)\ (\alpha + \beta) \otimes x = \alpha \otimes x \oplus \beta \otimes x,\ \forall \alpha, \beta \in K,\ x \in K;$$

$$3)\ \alpha \otimes (\beta \otimes x) = (\alpha \cdot \beta) \otimes x,\ \forall x \in X,\ \alpha, \beta \in K;$$

$$4)\ 1_K \otimes x = x,\ \forall x \in X,\ \text{where}\ 1_k\ \text{is the neuter element}$$

in relation to the operation multiplicatively marked in K.

In the context of linear spaces over arbitrary fields $R^3 = \{(x, y, z) : x, y, z \in R\}$ is a real linear (vector) space, which also allows the visualization of the concept of vector, which we define in this situation, if we take into account any system of axioms on which Euclidean geometry is based on.

Any ordered pair (A, B) of points in R^3 is called *oriented segment* of A origin, B extremity, sense from A to B and direction corresponding to the straight line AB. We will write $(A, B) = \overline{AB}$, and the length of any oriented segment of this kind will be indicated by $\|AB\|$. As the sense and direction form the *orientation* of any oriented segment, then an arbitrary oriented segment is characterized by orientation and length. In the set S of all the oriented segments from R^3 the following relation called *relation of equipolence* is considered: if $s_1, s_2 \in S$, then s_1 is *equipolent* with s_2, in notation $s_1 \sim s_2$ any time when s_1 and s_2 have the same length and orientation.

The defined relation of equipolence is an authentic relation of equivalence, being *reflexive* ($s \sim s,\ \forall s \in S$), symmetrical ($\forall s_1, s_2 \in S$ with

$s_1 \sim s_2 \Rightarrow s_2 \sim s_1$) and *transitive* ($\forall s_1, s_2, s_3 \in S$, so that $s_1 \sim s_2$ and $s_2 \sim s_3 \Rightarrow s_1 \sim s_3$).

Any class of equivalence (the set of all equipolent oriented segments with an oriented segment) is called *vector* in R^3.

Thus, the description of any vector $\overrightarrow{AB}$ is $\overrightarrow{AB} = \{\overline{CD} \in S : \overline{CD} \sim \overline{AB}\}$, the length $\left\|\overrightarrow{AB}\right\|$ (respectively the orientation) being the same with the one common to all oriented segments of the equivalence class that defines it. It is obvious that there are no "free vectors" and no "tied vectors", names used sometimes in current classes. If V is the set of vectors in R^3, the addition operation $+ : V \times V \to V$ is defined in the following manner : if $\overrightarrow{v_1}, \overrightarrow{v_2} \in V$ and P is an arbitrary point in R^3, the oriented segments $\overrightarrow{PA} \in \overrightarrow{v_1}$ and $\overrightarrow{PB} \in \overrightarrow{v_2}$ are considered and the $PACB$ parallelogram is constructed (see Figure 13):

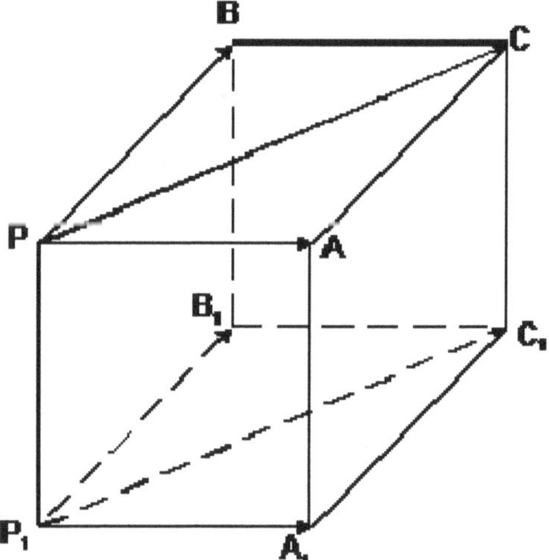

Figure 13: The parallelogram technique

By definition $\overrightarrow{v_1} + \overrightarrow{v_2} = \{\overline{TS} \in S : \overline{TS} \sim \overline{PC}\}$

The simple geometric arguments show that the definition is substantial (it is not dependent of the point P and of the chosen representatives for the vectors $\overrightarrow{v_1}$ and $\overrightarrow{v_2}$); indeed, for every

$P_1 \in R^3$, $\overline{P_1A_1} \in \vec{v_1}$ and $\overline{P_1B_1} \in \vec{v_2}$, by considering the $P_1A_1C_1B_1$ parallelogram, it is immediately noticed that $\overline{PC} \sim \overline{P_1C_1}$.

Another important operation defined on V is the *real scalar multiplication* $: R \times V \to V$ which attaches to any couple $(\alpha, \vec{v}) \in R \times V$ a vector $\overrightarrow{\alpha \cdot u}$ characterized by: $\left\| \alpha \cdot \vec{v} \right\| = |\alpha| \cdot \left\| v \right\|$, it has the same direction with the vector $\vec{v}$, the same sense for $\alpha \geq 0$ and contrary sense to the sense of vector $\vec{v}$ if $\alpha \leq 0$. Any two vectors $\vec{u_1}, \vec{u_2} \in V$ for which there exists $\beta \in R$ so that $\vec{u_1} = \beta \cdot \vec{u_2}$ or $\vec{u_2} = \beta \cdot \vec{u_1}$ are called *colinear*. We remind that by the term *axis* it is understood any straight line on which an origin, a length unit and a sense were established. In any situation of this kind, the vector with length 1 and orientation identical with the orientation of the axis, is called *the versor* of the axis.

If we mark with $\vec{i}, \vec{j}, \vec{k}$ the corresponding versors of the axes O*x*, O*y* respectively O*z* from a current system of axes, mutually orthogonal with the same origin O, then, taking into account the defined operation and the following representation, it follows that for every $A(x, y, z) \in R^3$, $\overline{OA} = x\vec{i} + y\vec{j} + z\vec{k}$

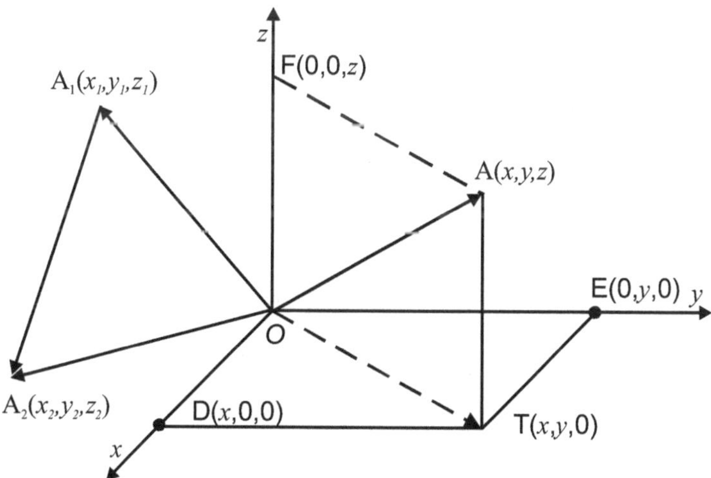

Figure 14: A projection of this 3-dimensional operation.

Indeed, $\overrightarrow{OA} = \overrightarrow{OT} + \overrightarrow{TA} = \overrightarrow{OE} + \overrightarrow{OD} + \overrightarrow{TA} = x\vec{i} + y\vec{j} + z\vec{k}$. For an arbitrary vector $\vec{u} = \overrightarrow{A_1 A_2} \in V$ with $A_1(x_1, y_1, z_1)$ and $A_2(x_2, y_2, z_2) \in R_3$ is obtained $\vec{u} = \overrightarrow{OA_2} - \overrightarrow{OA_1} = (x_2 - x_1)\vec{i} + (y_2 - y_1)\vec{j} + (z_2 - z_1)\vec{k}$. With these explanations the immediate properties of addition, respectively of scalar multiplication (that can also be directly deducted, from the definitions of these operations) are found:

1) $(\vec{v_1} + \vec{v_2}) + \vec{v_3} = \vec{v_1} + (\vec{v_2} + \vec{v_3})$, $\forall \vec{v_1}, \vec{v_2}, \vec{v_3} \in V$;

2) $\exists\, \vec{\theta}$ (the null vector made up of all the oriented segments whose origin coincides with the end), $\forall\, \vec{v} \in V : \vec{\theta} + \vec{v} = \vec{v} + \vec{\theta} = \vec{v}$;

3) $\forall\, \vec{v} \in V$, $\exists\, (-\vec{v}) \in V$ ($-\vec{v}$ is called the *contrary* of the vector $\vec{v}$, with the length $\left\| -\vec{v} \right\| = \left\| \vec{v} \right\|$, having the same direction with the vector $\vec{v}$ and contrary sense to vector $\vec{v}$) $: \vec{v} + (-\vec{v}) = (-\vec{v}) + \vec{v} = \vec{\theta}$;

4) $\vec{v_1} + \vec{v_2} = \vec{v_2} + \vec{v_1}$, $\forall \vec{v_1}, \vec{v_2} \in V$;

1) – 4) $\Rightarrow$ $(V, + : V \times V \to V)$ Abelian (commutative) group

5) $\alpha \cdot (\vec{v_1} + \vec{v_2}) = \alpha \cdot \vec{v_1} + \alpha \cdot \vec{v_2}$, $\forall \vec{v_1}, \vec{v_2} \in V$, $\alpha \in R$;

6) $(\alpha + \beta) \cdot \vec{v} = \alpha\vec{v} + \beta\vec{v}$, $\forall \vec{v} \in V$, $\alpha, \beta \in R$;

7) $\alpha \cdot (\beta \cdot \vec{v}) = (\alpha\beta) \cdot \vec{v} = \beta \cdot (\alpha \cdot \vec{v})$, $\forall \vec{v} \in V$, $\alpha, \beta \in R$;

8) $1 \cdot \vec{v} = \vec{v}$, $\forall \vec{v} \in V$

1)–8) $\Rightarrow$ $(V, + : V \times V \to V, \cdot R \times V \to V)$ *real linear* (vectorial) space. Important linear subspaces: $V_1 = \{x\vec{i} : x \in R\}$ - the real linear subspace of vectors from any versor axis $\vec{i}$, $V_2 = \{x\vec{i} + y\vec{j} : x, y \in R\}$ - the real linear subspace of vectors from any plane established by the support straight lines of two distinct axes of versors $\vec{i}$ respectively $\vec{j}$.

By definition, *the scalar product* of two arbitrary vectors $\vec{v_1}, \vec{v_2} \in V$ is the real number $\langle \vec{v_1}, \vec{v_2} \rangle = \|\vec{v_1}\| \cdot \|\vec{v_2}\| \cdot \cos(\vec{v_1}, \vec{v_2})$, and the application $\langle \cdot, \cdot \rangle : V \times V \to R$ satisfies the following characteristic conditions:

6) $\langle \vec{v}, \vec{v} \rangle = \|\vec{v}\|^2 \geq 0, \ \forall \vec{v} \in V, \ \langle \vec{v}, \vec{v} \rangle = 0 \Leftrightarrow \vec{v} = \vec{\theta}$;

7) $\langle \vec{v_1}, \vec{v_1} \rangle = \langle \vec{v_2}, \vec{v_1} \rangle, \ \forall \vec{v_1}, \vec{v_2} \in V$;

8) $\langle \alpha \vec{v_1} + \beta \vec{v_2}, \vec{v_3} \rangle = \alpha \langle \vec{v_1}, \vec{v_2} \rangle + \beta \langle \vec{v_1}, \vec{v_3} \rangle, \ \forall \vec{v_1}, \vec{v_2}, \vec{v_3} \in V, \ \alpha, \beta \in R$

Furthermore,

9) $\left| \langle \vec{v_1}, \vec{v_2} \rangle \right| \leq \|\vec{v_1}\| \cdot \|\vec{v_2}\|, \ \forall \vec{v_1}, \vec{v_2} \in V$ and

10) $\left| \langle \vec{v_1}, \vec{v_2} \rangle \right| \leq \dfrac{\|\vec{v_1}\|^2 + \|\vec{v_2}\|^2}{2}, \ \forall \vec{v_1}, \vec{v_2} \in V$.

Two vectors $\vec{v_1}, \vec{v_2} \in V$ are *orthogonal*, in notation, $\vec{v_1} \perp \vec{v_2}$ if $\langle \vec{v_1}, \vec{v_2} \rangle = 0$. For any vectors in R^3 $\vec{v_1} = x_1 \vec{i} + y_1 \vec{j} + z_1 \vec{k}$ and $\vec{v_2} = x_2 \vec{i} + y_2 \vec{j} + z_2 \vec{k}$, $\langle \vec{v_1}, \vec{v_2} \rangle = x_1 x_2 + y_1 y_2 + z_1 z_2$ because $\vec{i} \perp \vec{j} \perp \vec{k} \perp \vec{i}$.

If $A_1(x_1, y_1, z_1), A_2(x_2, y_2, z_2) \subset R^3$ are any elements, then

$$\|A_1 A_2\| = \left\|\overrightarrow{A_1 A_2}\right\| = \sqrt{\left\|\overrightarrow{A_1 A_2}\right\|^2} = \sqrt{\langle \overrightarrow{A_1 A_2}, \overrightarrow{A_1 A_2} \rangle} = \sqrt{(x_2 - x_1)^2 + (y_2 - y_1)^2 + (z_2 - z_1)^2}$$

The explanations presented in this chapter can be transferred "mutatis mutandis" in any $n \in N^*$ dimensional Euclidean space.

CHAPTER 4

The Fundamental Trigonometric Functions and the Customary Inversive Restrictions

Abstract: This part of this book is devoted to the fundamental trigonometric functions examined together with their inverted restrictions, following the main directions of studies specified at the end of Chapter 2.

Keywords: Angle, measure, trigonometric circle, sine, cosine, tangent, cotangent function, their corresponding inverted functions, graphics, properties.

4.1 INTRODUCTION

We remind that by "*angle*" is understood any couple of semi-straights with the same origin. The side from which the angle is measured is called *the initial side* and the one where it ends is called *the final side*. By convention, all clock-wise measured angles get in front of their measure the "-" sign, and the others the "+" sign (see Figure 15).

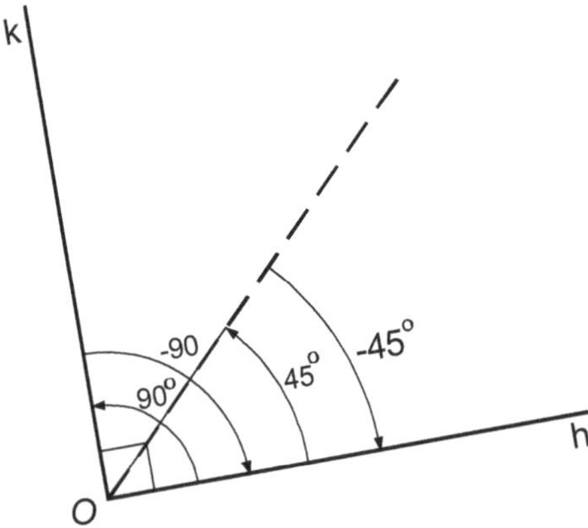

Figure 15: The measure of the angles.

Any circle having the radius length equal with the chosen measurement unit is called *trigonometric circle*. An angle equals 1 radian (in short 1 rad), if the trigonometric circle centered in the tip of the angle includes the sides of this arc with the length equal to 1, that is having the length equal to that of the circle radius.. (see Figure 16).

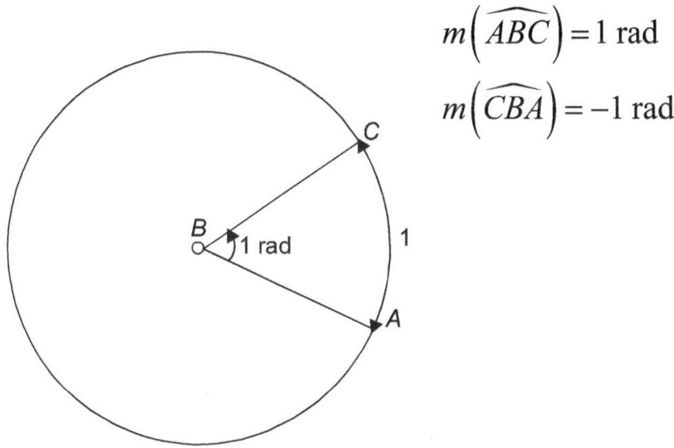

$$m\left(\widehat{ABC}\right) = 1 \text{ rad}$$

$$m\left(\widehat{CBA}\right) = -1 \text{ rad}$$

Figure 16: The possibility of measure by rads.

Generally speaking, a $\widehat{TVS}$ angle equals y radians if

$$\frac{aria\left(TVS\right)}{aria\,D\left(V;1\right)} = \frac{y}{2\pi} \text{ , where } D\left(V;1\right) \text{ is the disc centered in } V \text{ of radius 1,}$$

and $\left(TVS\right)$ the inducted circular sector (see Figure 17).

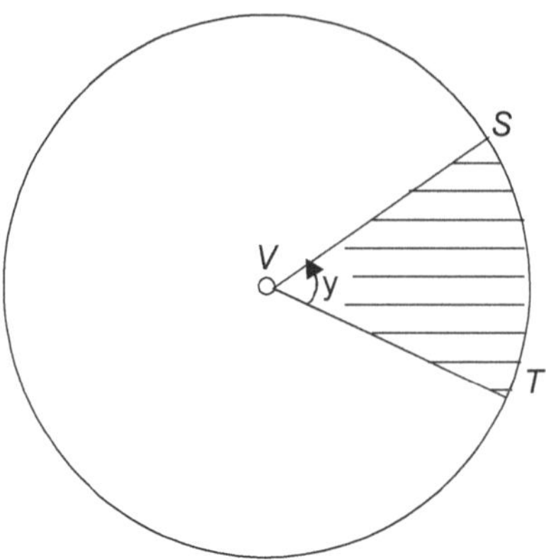

Figure 17: Its continuity

The replacement of the number 2π by another number was proved to be inefficient, following at least the numerical methods. The correspondence between the measure x^0 of an angle expressed in sexagesimal degrees than the corresponding one in radians is immediate illustrated by

$$\pi\,\frac{x^{\circ}}{180^0}\,(rad).$$

In the following pages we will use the results of the Chapters I - III projected to the real functions of real arguments.

4.2. THE SINE FUNCTION

It is marked by $\sin : R \to [-1,1]$ and defined for any angle with its measure α expressed in radians as *being the second coordinate of the point of intersection between the final side of the angle and the trigonometric circle centered on the tip of the angle, in the system of coordinates in which the* Ox *axis is the support straight of the initial side, and the* Oy *axis the perpendiculary axis on* Ox, through the tip of the angle, both being oriented as follows (see Figure 18).

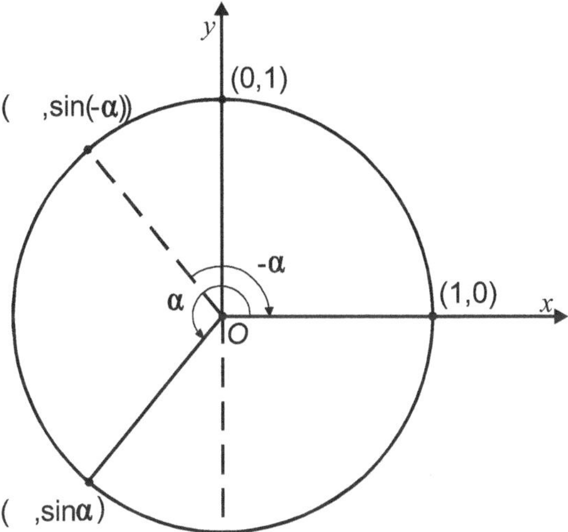

Figure 18: The simple geometrical definition of $\sin : R \rightarrow \left[-1,1\right]$

The precedent introduction of the sine functon shows that $\left|\sin x\right| \leq \left|x\right|$ for any $x \in R$ and easily leads to the following properties.

    1. The peridiocity

$\sin\left(2k\pi + \alpha\right) = \sin\alpha, \forall \alpha \in R, \forall k \in Z.$ Thus, the sine function is periodic, is of principal period 2π, and the graphic representation is obtained successively, by adequate translations from the one corresponding to the interval $\left[0, 2\pi\right]$.

  2. The imparity

$\sin\left(-\alpha\right) = -\sin\alpha, \forall \alpha \in R,$ hence the sine function is odd, which implies that the representation of the graphic be symmetrical to the origin of the coordinate system in which it is realised.

  3.   *The representation of the graphic*

x	0	$\frac{\pi}{2}$	π	$\frac{3\pi}{2}$	2π
$\sin x$	0	1	0	-1	0

$,\sin k\pi = 0, \forall k \in Z$ (see Figure 19).

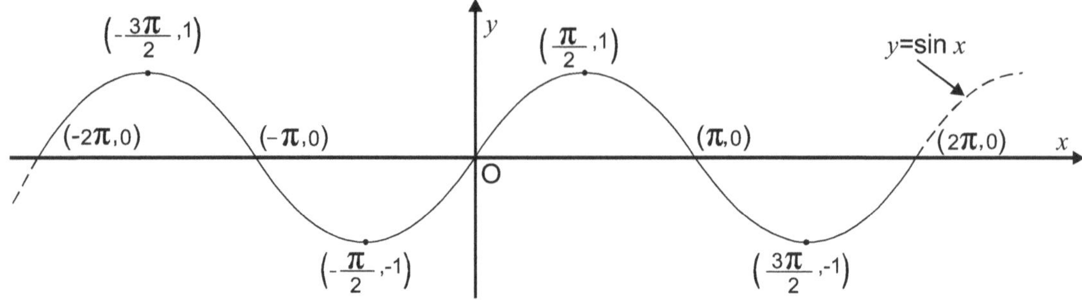

Figure 19: *The sine graphic oscillation.*

4. *The inversiveness*

We can notice even from the graphic representation that the sine function is not bijective (isn't injective also), because there are parallels to Ox axis, which crosses the graphic representation in an infinity of points. For example, any straight $y = c$ with $c \in [-1,1]$, but allows bijective restrictions, that are inversive. This kind of restriction is:

$$\sin\Big/_{\left[-\frac{\pi}{2},\frac{\pi}{2}\right]} : \left[-\frac{\pi}{2}, \frac{\pi}{2}\right] \rightarrow [-1,1]$$

whose inverse mapping is marked by

$$\arcsin = \left(\sin\Big/_{\left[-\frac{\pi}{2},\frac{\pi}{2}\right]}\right)^{-1} : [-1,1] \rightarrow \left[-\frac{\pi}{2}, \frac{\pi}{2}\right].$$

Thus, $y = \arcsin x \Leftrightarrow \begin{cases} |x| \leq 1 \\ y \in \left[-\dfrac{\pi}{2}, \dfrac{\pi}{2}\right], \\ \sin y = x \end{cases}$ mentioning that *only* the inversion of this restriction is marked like that, the graphic representation of the function $x \rightarrow \arcsin x$ is obtained through the graphic representation of the restriction $\sin\big/_{\left[-\frac{\pi}{2},\frac{\pi}{2}\right]}$ by symmetry to the first bisectrix

(see Figure 20.)

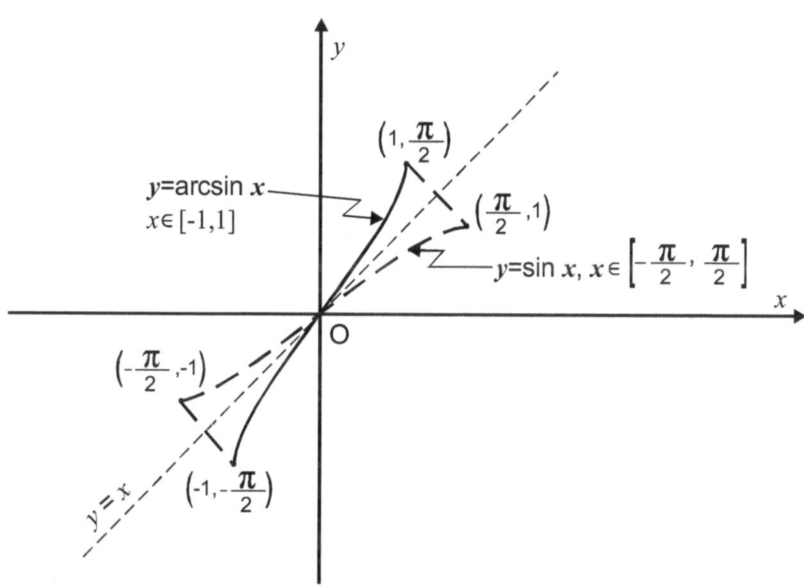

Figure 20: The graphic representation of the main inverted restriction on $\left[-\dfrac{\pi}{2},\dfrac{\pi}{2}\right]$.

5. The sign

$$\sin u \leq 0 \Leftrightarrow u \in \bigcup_{k \in Z}\left[2k\pi + \pi, 2k\pi + 2\pi\right];$$

$$\sin u < 0 \Leftrightarrow u \in \bigcup_{k \in Z}\left(2k\pi + \pi, 2k\pi + 2\pi\right);$$

$$\sin u \geq 0 \Leftrightarrow u \in \bigcup_{k \in Z}\left[2k\pi, 2k\pi + \pi\right];$$

$$\sin u > 0 \Leftrightarrow u \in \bigcup_{k \in Z}\left(2k\pi, 2k\pi + \pi\right).$$

6. The monotony

Observing the graphic representation it can be noticed that the sine function is strictly decreasing on every interval

$$\left[2k\pi+\frac{\pi}{2},2k\pi+\frac{3\pi}{2}\right](k\in Z),$$ strictly increasing on any of the

intervals of the type $\left[2k\pi-\frac{\pi}{2},2k\pi+\frac{\pi}{2}\right](k\in Z)$ and the following

relations are obvious:

$$\sin u=\gamma\in\left[-1,1\right]\Leftrightarrow u=k\pi+\left(-1\right)^{k}\arcsin\gamma,k\in Z;$$

$$\sin u=\sin v\Leftrightarrow u=k\pi+\left(-1\right)^{k}v,k\in Z.$$

We write $\operatorname{cosec}\alpha=\dfrac{1}{\sin\alpha},\forall\alpha\in R\setminus\{k\pi:k\in Z\}.$

7. The boundedness

$\left|\sin x\right|\le 1,\forall x\in R,$ thus the sine function is bounded

on *R*.

8. The continuity

Sine is a continuous function on *R*, thus it also has the Darboux property on the real   set.

9. The convexity (concavity).

Sine is a strictly convex function on each interval $\left[2k\pi-\pi,2k\pi\right](k\in Z)$ and strictly concave on any interval $\left[2k\pi,2k\pi+\pi\right],k\in Z.$

Following the presentation model for the sine function we will present the cosine, tangent and cotangent trigonometric function along with the corresponding prioritary inversive restrictions.

4.3. THE COSINE FUNCTION

Marked by $\cos : R \to [-1,1]$ *and defined for every angle of measure* $\alpha \in R$ *radians as the first coordinate of the point of intersection between the final side of the angle and the trigonometric circle centered in its tip,* in the same system of coordinates

mentioned for the sine function (see Figure 21).

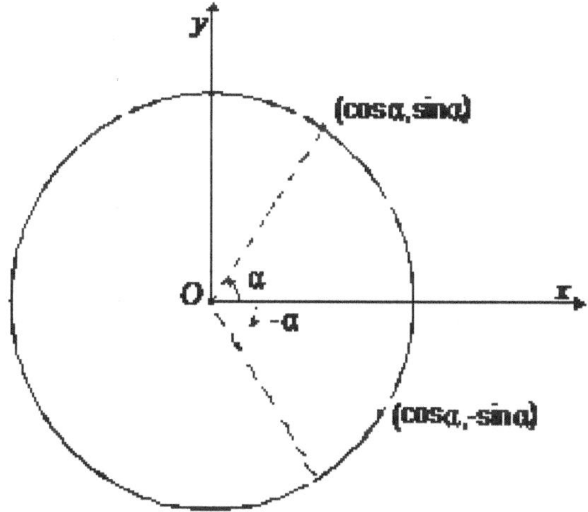

Figure 21: The geometrical definition .

From the definition one obtains immediately the following properties.

1. *The peridiocity*

$\cos(2k\pi + \alpha) = \cos(\alpha), \forall \alpha \in R, k \in Z,$　principal period being 2π.

2. *The parity*

$\cos(-\alpha) = \cos\alpha, \forall \alpha \in R,$　　thus　　the　　graphic representation of the cosine function allows Oy as symmetry axis.

3. *The graphic representation*

The periodicity combined with the following values leads to: $\cos k\pi = (-1)^k, \forall k \in Z$ (see Figure 22).

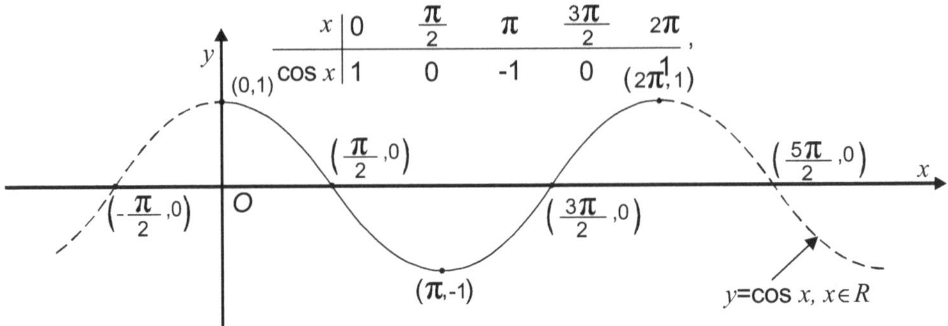

Figure 22: The cosine graphic oscillation

4. *The inversiveness*

Like in the case of the sine function, the cosine function isn't bijective, not being injective either, but allows some inversive restrictions.

The inverse mapping of the restriction $\cos/_{[0,\pi]} : [0,\pi] \to [-1,1]$ and *only* this one is marked with $\left(\cos/_{[0,\pi]}\right)^{-1} = \arccos : [-1,1] \to [0,\pi].$

Thus, $y = \arccos x \Leftrightarrow \begin{cases} |x| \le 1 \\ y \in [0,\pi], \\ \cos y = x \end{cases}$ and the graphic representation is the following (see Figure 23).

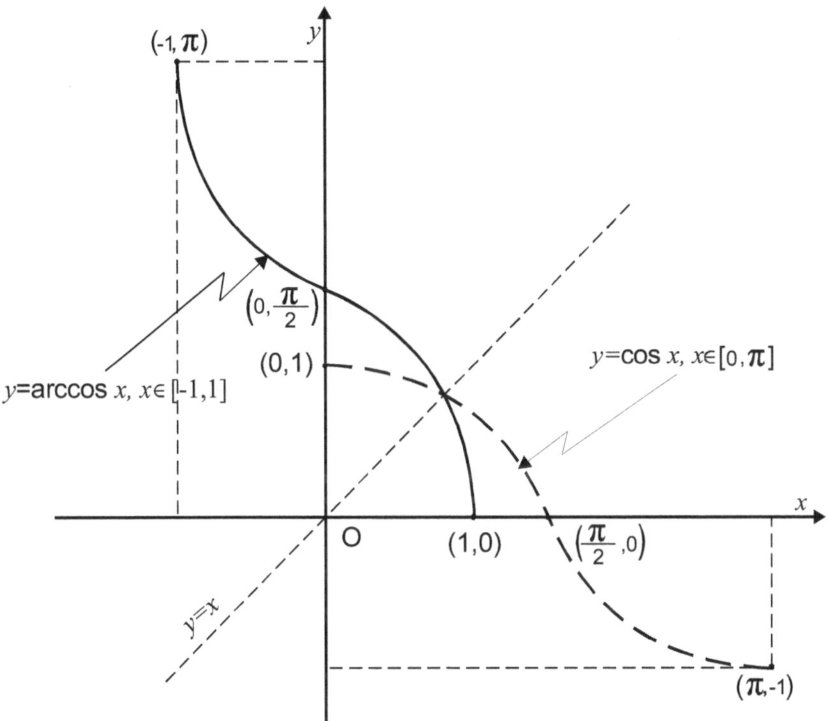

Figure 23: Its inverted graphic's on $[0, \pi]$

5. *The sign*

$$\cos u \leq 0 \Leftrightarrow u \in \bigcup_{k \in Z}\left[2k\pi + \frac{\pi}{2}, 2k\pi + \frac{3\pi}{2}\right];$$

$$\cos u < 0 \Leftrightarrow u \in \bigcup_{k \in Z}\left(2k\pi + \frac{\pi}{2}, 2k\pi + \frac{3\pi}{2}\right);$$

$$\cos u \geq 0 \Leftrightarrow u \in \bigcup_{k \in Z}\left[2k\pi - \frac{\pi}{2}, 2k\pi, 2k\pi + \frac{\pi}{2}\right];$$

$$\cos u > 0 \Leftrightarrow u \in \bigcup_{k \in Z}\left(2k\pi - \frac{\pi}{2}, 2k\pi + \frac{\pi}{2}\right).$$

6. *The monotony*

Observing the graphic representation it can be noticed that the cosine function is strictly decreasing on every interval $[2k\pi, 2k\pi + \pi]\,(k \in Z)$, strictly increasing on any of the intervals of the type $[2k\pi + \pi, 2k\pi + 2\pi]$, $\cos u = \gamma \in [-1,1]$

$$\Leftrightarrow u = 2k\pi \pm \arccos \gamma, k \in Z$$

and $\cos u = \cos v \Leftrightarrow u = 2k\pi \pm v,\ k \in Z.$

By conventions,

$$\frac{1}{\cos \alpha} = \sec \alpha, \forall \alpha \in R \setminus \left\{ 2k\pi \pm \frac{\pi}{2} : k \in Z \right\}$$

The definitions of the sine and cosine function, together with Pythagoras' theorem conveniently applied or using the vectors in R^3 presented in Chapter III, motivate also the validity of the basic relation $\sin^2 u + \cos^2 u = 1,\ \forall u \in R.$ The surjectivity of the sine and cosine functions, together with the precedent equality, leads to the conclusion that any time $\alpha^2 + \beta^2 = 1\,(\alpha, \beta \in R)$, there exists $\theta \in [0, 2\pi]$ so that $\alpha = \sin \theta$ and $\beta = \cos \theta$ or $\alpha = \cos \theta$ and $\beta = \sin \theta$.

7. The boundedness

$|\cos x| \leq 1, \forall x \in R,$ results from the definition, which confirms the boundedness of the cosine function on R.

8. The continuity

The cosine function is continuous on the real number set, fact that supports the validity of Darboux property on R.

9. The convexity (concavity).

Cosine is a *strictly convex* function on each interval $\left[2k\pi + \dfrac{\pi}{2}, 2k\pi + \dfrac{3\pi}{2} \right](k \in Z)$ and *strictly concave* on any interval $\left[2k\pi - \dfrac{\pi}{2}, 2k\pi + \dfrac{\pi}{2} \right], k \in Z.$

4.4. THE TANGENT FUNCTION

Is marked by $\mathrm{tg}: R \setminus \left\{ k\pi + \dfrac{\pi}{2} : k \in Z \right\} \to R$ and $\mathrm{tg}\,\alpha$ *is defined as the second coordonate of the point of intersection between the support straight of the final side for each measure angle* $\alpha \in R \setminus \left\{ k\pi + \dfrac{\pi}{2} : k \in Z \right\}$ *radians and the geometric tangent to the trigonometric circle centered on the tip of the angle related to the point of coordinates* $(1,0),$ the system of coordinates being chosen in an identical manner to the definition of sine and cosine functions (see Figure 24).

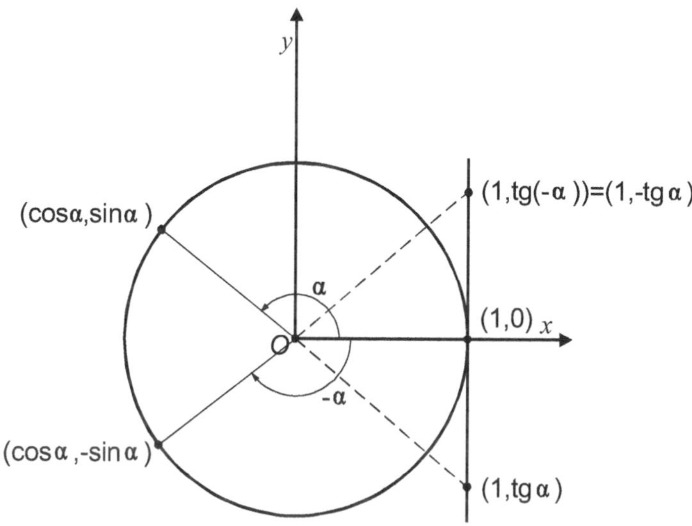

Figure 24: The tangent geometrical definition

It is obvious that any measure angle $\beta = k\pi + \dfrac{\pi}{2}, k \in Z$ doesn't allow a tangent , because the support straight of the final side coincides with the support of the Oy axis. It can be noticed also that

$$|x| \le |\mathrm{tg}\,x|, \forall x \in R \setminus \left\{ k\pi + \frac{\pi}{2} : k \in Z \right\}.$$

1. *The periodicity*

$$\operatorname{tg}(q\pi + \alpha) = \operatorname{tg}\alpha, \forall \alpha \in R \setminus \left\{ k\pi + \frac{\pi}{2} : k \in Z \right\}, q \in Z, \text{ with}$$

the principal period π.

2. *The imparity*

$$\operatorname{tg}(-\alpha) = -\operatorname{tg}\alpha, \forall \alpha \in R \setminus \left\{ k\pi + \frac{\pi}{2} : k \in Z \right\}, \text{ thus the}$$

graphic representation of the tangent function

allows as symmetry center the origin

of the coordination axes.

3. *The graphic representation*

Periodicity and the following sequence of values (see Figure 25).

x	$-\dfrac{\pi}{2}$	$-\dfrac{\pi}{4}$	0	$\dfrac{\pi}{4}$	$\dfrac{\pi}{2}$
$\operatorname{tg} x$	$\|$	-1	0	$\|$	1

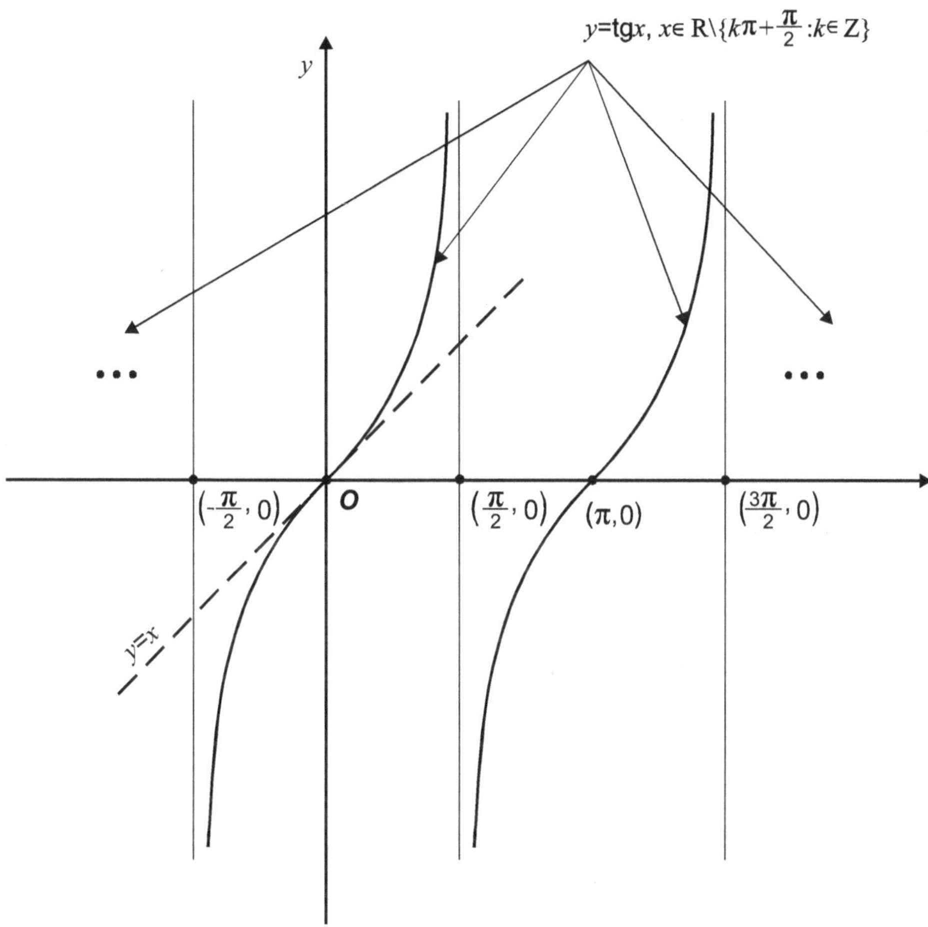

Figure 25: Its graphic representation

4. *The inversiveness*

The absence of the injectivity property doesn't allow the existence of the inversion on the set of definition, but the tangent function allows inversive restrictions.

This kind of restriction is the bijective function

$$\mathrm{tg}\Big/_{\left(-\frac{\pi}{2},\frac{\pi}{2}\right)}:\left(-\frac{\pi}{2},\frac{\pi}{2}\right)\to R$$

with the corresponding inverse mapping marked in *unique* manner

$$\left(tg \Big/_{\left(-\frac{\pi}{2},\frac{\pi}{2}\right)} \right)^{-1} = \arctg : R \to \left(-\frac{\pi}{2},\frac{\pi}{2} \right).$$

Both bijective function can be generalised by replacing the interval $\left(-\dfrac{\pi}{2},\dfrac{\pi}{2} \right)$ with an arbitrary, uncommon, bounded interval $\left(k\pi - \dfrac{\pi}{2}, k\pi + \dfrac{\pi}{2} \right)(k \in Z)$ where the bijectivity is kept. This fact is expressed, in the language of cardinals mentioned at the paragraph 1.2. of the first chapter, by the property that the real number set is *equipotent* (it realises a bijective correspondence) with any interval I of this kind.

$$\text{Thus, } \ y = \arctg x \Leftrightarrow \begin{cases} y \in \left(-\dfrac{\pi}{2},\dfrac{\pi}{2} \right), \\ tg\, y = x \end{cases} \text{, and the graphic}$$

representations obtained by symmetry to the first bisectrix of the graphic representation of the considered restriction (see Figure 26).

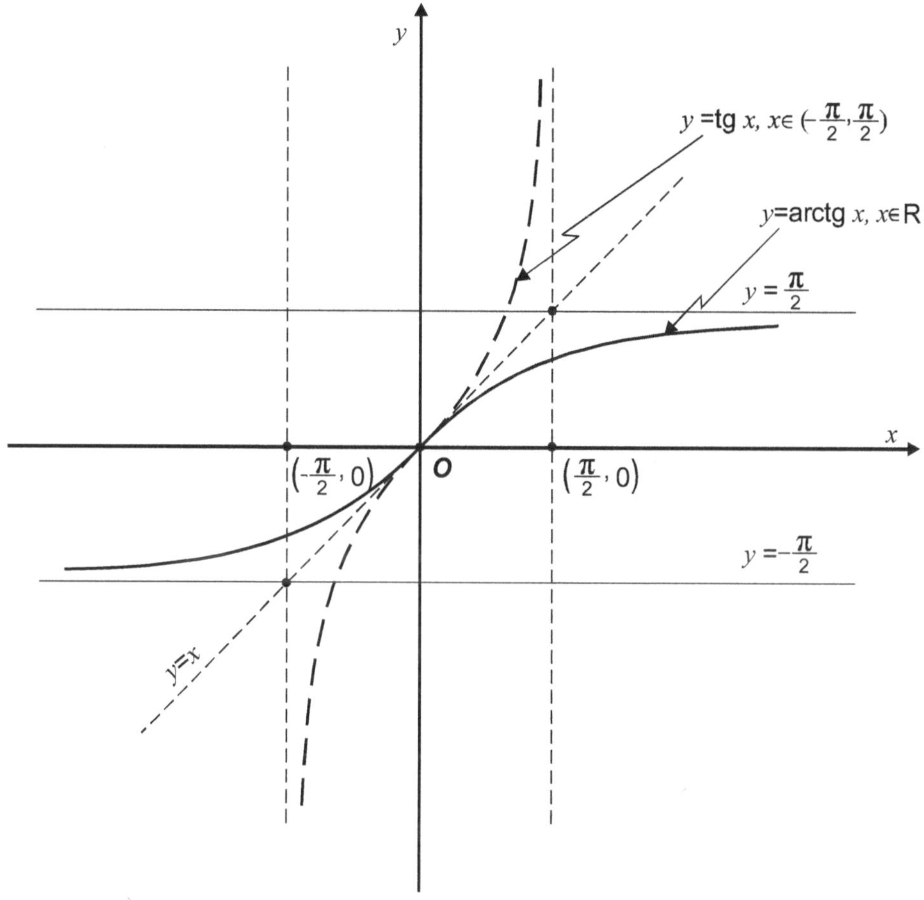

Figure 26: The graphic representation of the main inverted restriction

5. *The sign*

$$\mathrm{tg}\, u \leq 0 \Leftrightarrow u \subset \bigcup_{k \in Z} \left(k\pi - \frac{\pi}{2}, k\pi \right];$$

$$\mathrm{tg}\, u < 0 \Leftrightarrow u \in \bigcup_{k \in Z} \left(k\pi - \frac{\pi}{2}, k\pi \right);$$

$$\mathrm{tg}\, u \geq 0 \Leftrightarrow u \in \bigcup_{k \in Z} \left[k\pi, k\pi + \frac{\pi}{2} \right);$$

$$\mathrm{tg}\, u > 0 \Leftrightarrow u \in \bigcup_{k \in Z} \left(k\pi, k\pi + \frac{\pi}{2} \right).$$

6. *The monotony*

Observing the graphic representation it can be noticed that the tangent function is strictly increasing on any interval $\left(k\pi - \dfrac{\pi}{2}, k\pi + \dfrac{\pi}{2} \right), k \in Z$

and

$$\operatorname{tg} u = \gamma \in R \Leftrightarrow u = k\pi + \operatorname{arctg} \gamma, k \in Z,$$

$$\operatorname{tg} u = \operatorname{tg} v \Leftrightarrow u = k\pi + v, k \in Z.$$

Furthermore, from the next geometrical projection by its definition, it follows that we have (see Figure 27).

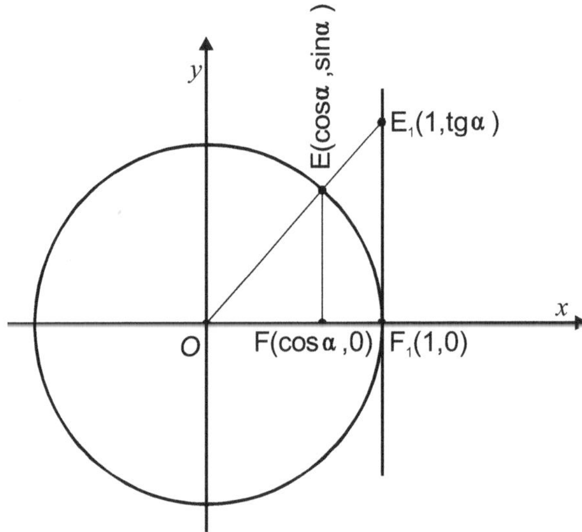

Figure 27: Sine and cosine together with tangent.

Obviously, the similarity relation between ΔEOF and $\Delta E_1 OF_1$ generated for any angle of measure $\alpha \in R \setminus \left\{ k\pi + \dfrac{\pi}{2} : k \in Z \right\}$ radians implies that

$$\operatorname{tg} \alpha = \frac{\sin \alpha}{\cos \alpha}.$$

7. The unboundedness

From the graphic representation it can be noticed the *unboundedness* of the tangent function in any neighborhood of each point $k\pi + \dfrac{\pi}{2}, k \in Z$ on which the function is defined, so the tangent is unbounded on the set of definition as well.

8. The continuity

The tangent function is continuous on $R \setminus \left\{ k\pi + \dfrac{\pi}{2} : k \in Z \right\}$, property which is validated also by the relation $\operatorname{tg}\alpha = \dfrac{\sin\alpha}{\cos\alpha}$ combined with the definition set of the tangent functionand the continuity of the sine and cosine function on R mantained by the algebraic operations that define the anounced relation.

9. The convexity (concavity).

the graphic representation warrants the *strictly convexity* of the tangent function on each interval $\left[k\pi, k\pi + \dfrac{\pi}{2} \right)(k \in Z)$ and the *strictly concavity* of the same function on $\left(k\pi - \dfrac{\pi}{2}, k\pi \right], \forall k \in Z.$

4.5. THE COTANGENT FUNCTION

It is marked with $\operatorname{ctg} : R \setminus \{ k\pi : k \in Z \} \to R,$

and $\operatorname{ctg}\alpha$ defined for every measure angle $\alpha \in R \setminus \{ k\pi : k \in Z \}$ radians as *the first coordinate of the point of intersection between the support of the final side of the angle and the geometric tangent to the trigonometric circle centered in the tip of the angle in the point of coordinates* $(0,1)$, the system of coordinates being everytime chosen in a similar way to that used for the introduction of the sine, cosine, tangent (see Figure 28).

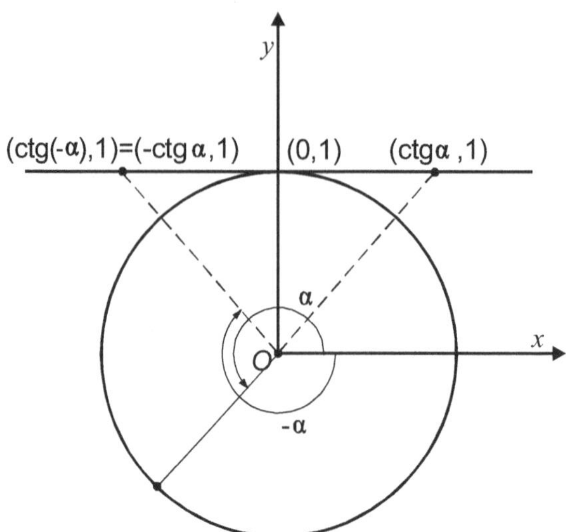

Figure 28: The cotangent function defined by geometrical arguments

It is obvious that there isn't $\operatorname{ctg}\beta$ for any measure angle $\beta = k\pi, k \in Z$ because the support straights of the final sides corresponding to all of these angles coincide with the support of Ox axis.

    1. *The periodicity*

$$\operatorname{ctg}(q\pi + \alpha) = \operatorname{ctg}\alpha, \forall \alpha \in R \setminus \{k\pi : k \in Z\}, q \in Z,\quad \text{with}$$

the principal period π.

    2. *The imparity*

$$\operatorname{ctg}(-\alpha) = -\operatorname{ctg}\alpha, \forall \alpha \in R \setminus \{k\pi : k \in Z\},\quad \text{so}\quad \text{the}$$

common origin of the coordination axes in any Cartesian system of identification of the graphic projected in R^2 is the center of symmetry for the graphic representation.

    3. *The graphic representation*

The precedent properties together with the immediate values lead to the next indicated graphic representation (see Figure 29.)

x	0	$\frac{\pi}{4}$	$\frac{\pi}{2}$	$\frac{3\pi}{4}$	π
ctg x	l	1	0	-1	l

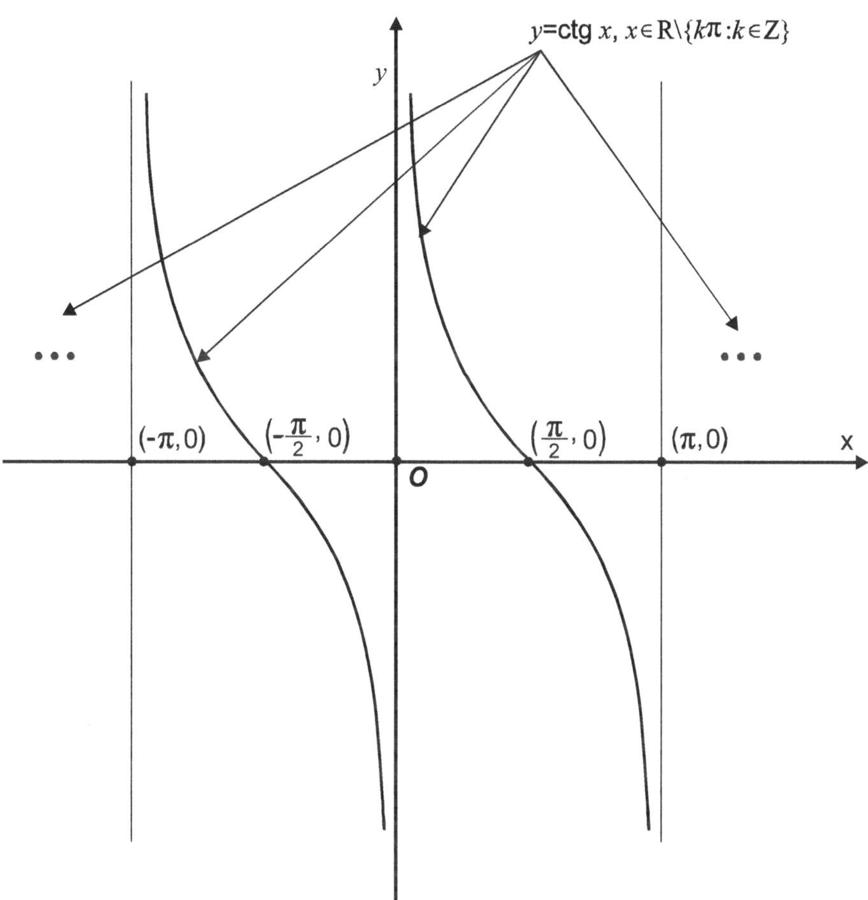

Figure 29: Its graphic representation.

4. *The inversiveness*

Like in the case of the above presented functions, the cotangent function is not inversive on the definition set due to the fact that it isn't injective, but allows inversive restrictions.

For example, *only* the inverse mapping of the bijective restriction $\mathrm{ctg}/_{(0,\pi)} : (0,\pi) \to R$ is marked by $\mathrm{arcctg} = \left(\mathrm{ctg}/_{(0,\pi)}\right)^{-1} : R \to (0,\pi)$ and can be graphically represented by symmetry to the first bisectrix of the graphic

representation of the function $\mathrm{ctg}/_{(0,\pi)}$ (see Figure 30).

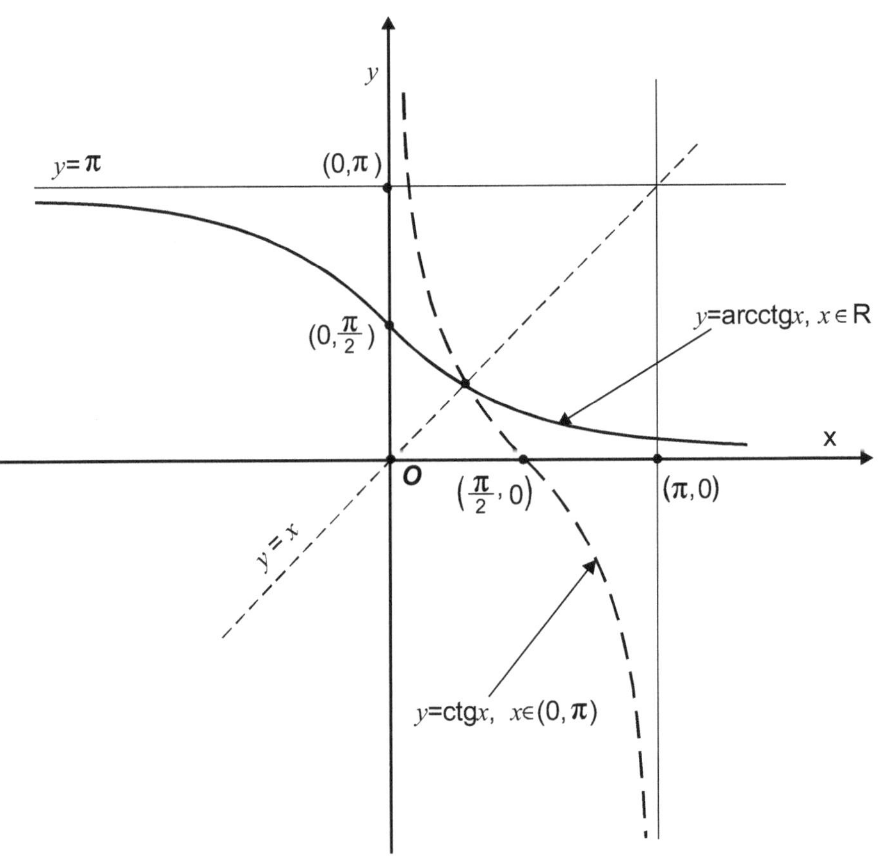

Figure 30: The graphic of the main inverted restriction.

Thus, $y = \operatorname{arcctg} x \Leftrightarrow \begin{cases} y \in (0, \pi) \\ \operatorname{ctg} y = x \end{cases}$.

5. *The sign*

$$\operatorname{ctg} u \leq 0 \Leftrightarrow u \in \bigcup_{k \in Z} \left[k\pi + \frac{\pi}{2}, k\pi + \pi \right);$$

$$\operatorname{ctg} u < 0 \Leftrightarrow u \in \bigcup_{k \in Z} \left(k\pi + \frac{\pi}{2}, k\pi + \pi \right);$$

$$\operatorname{ctg} u \geq 0 \Leftrightarrow u \in \bigcup_{k \in Z} \left(k\pi, k\pi + \frac{\pi}{2} \right];$$

$$\operatorname{ctg} u > 0 \Leftrightarrow u \in \bigcup_{k \in Z} \left(k\pi, k\pi + \frac{\pi}{2} \right).$$

6. *The monotony*

Observing the graphic representation it can be noticed that the cotangent function is strictly decreasing on each interval of the type $(k\pi, k\pi + \pi), k \in Z$ and

$$\operatorname{ctg} u = \gamma \in R \Leftrightarrow u = k\pi + \operatorname{arcctg} \gamma, k \in Z,$$
$$\operatorname{ctg} u = \operatorname{ctg} v \Leftrightarrow u = k\pi + v, k \in Z,$$

and the similarity relation $\Delta TOS \sim \Delta T_1 OS_1$ easily extended to any measure angle $\alpha \in R \setminus \{k\pi : k \in Z\}$ radians, leads to $\operatorname{ctg} \alpha = \dfrac{\cos \alpha}{\sin \alpha}$ (see Figure 31).

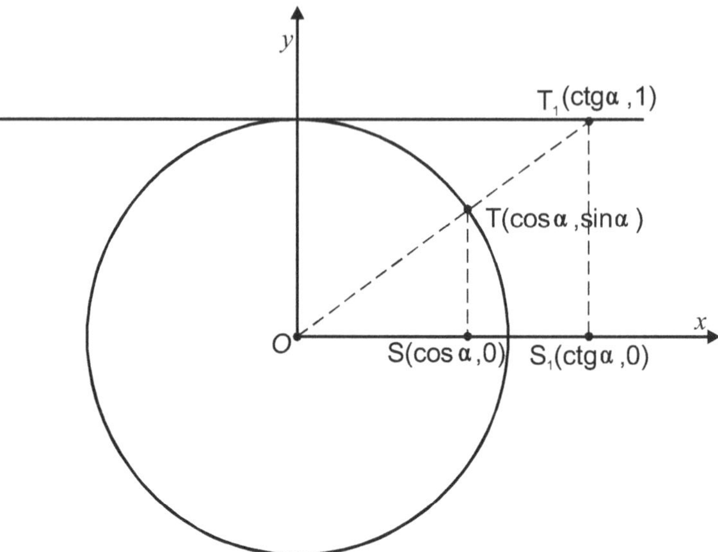

Figure 31: Geometrical motivation

7. *The unboundedness*

Also by observing the graphic representation the unboundedness of the cotangent function on any neighborhood from the definition set of each point $k\pi, k \in Z,$ is noticed, fact that generates the unboundedness of this function on the set $R \setminus \{k\pi : k \in Z\}.$

8. *The continuity*

Similar reasons that prove the continuity of the tangent function justifies the continuity of the cotangent function on the definition set.

9. The convexity (concavity).

The cotangent function is *strictly convex* in any interval $\left(k\pi, k\pi + \dfrac{\pi}{2} \right] (k \in Z)$ and *strictly concave* on each interval $\left[k\pi + \dfrac{\pi}{2}, k\pi + \pi \right), k \in Z.$

The previous statements show that the values of the trigonometric functions for any sharp angle $\widehat{ABC}$, from an arbitrary non-degenerated

rectangular triangle is calculated like this:

$$\sin\left(\widehat{ABC}\right) = \frac{\|AC\|}{\|BC\|}, \cos\left(\widehat{ABC}\right) = \frac{\|AB\|}{\|BC\|},$$

$$\operatorname{tg}\left(\widehat{ABC}\right) = \frac{\|AC\|}{\|AB\|}, \operatorname{ctg}\left(\widehat{ABC}\right) = \frac{\|AB\|}{\|AC\|}.$$

(see Figure 32).

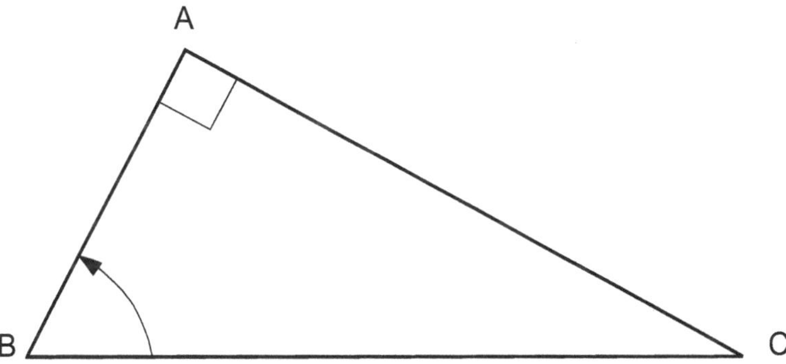

Figure 32: A projection in the plain

where the function $\|\cdot\|$ represents the length attached to the segments.

<h1>CHAPTER 5</h1>

Applications

Abstract: This section is for the essential algebraic connections related to the trigonometric functions introduced in the precedent chapters and for the immediate implications and applications in the Euclidean geometry.

Keywords: Trigonometric formulae, calculus by identities, inequalities, equations, inequations and their specified solutions, complex numbers' plane projection and geometrical examples.

5.1. TRIGONOMETRIC FORMULAE

Firstly we remind the basic property

$$\boxed{\sin^2 \alpha + \cos^2 \alpha = 1, \ \forall \alpha \in R} \tag{0}$$

established in the precedent chapter, which proves that:

$$\boxed{1 + \mathrm{tg}^2 \alpha = \sec^2 \alpha, \ \forall \alpha \subset R \setminus \{k\pi + \frac{\pi}{2} : k \subset Z\}}$$

and

$$\boxed{1 + \mathrm{ctg}^2 \beta = \operatorname{cosec}^2 \beta, \ \forall \beta \in R \setminus \{k\pi : k \in Z\}}$$

Let $\alpha, \beta \in R$ be the measurements in radians of the angles between the Ox axis and the vectors $\vec{u} = (\cos\alpha)\vec{i} + (\sin\alpha)\vec{j}$, $\vec{v} = (\cos\beta)\vec{i} + (\sin\beta)\vec{j}$ respectively (see Figure 33).

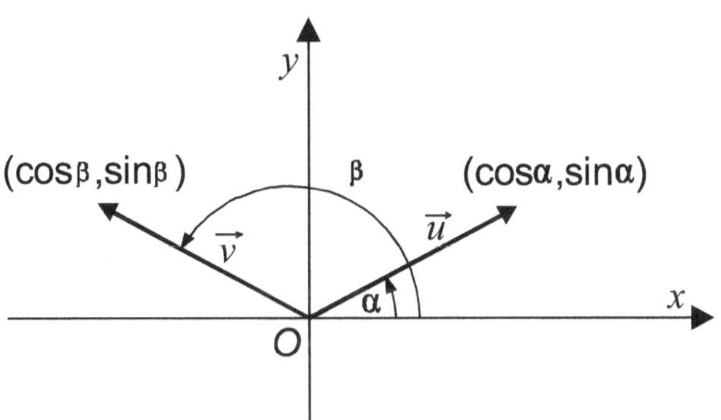

Figure 33: Vectors generated by sine and cosine

Following the considerations of Chapter III, the scalar product of the vectors $\vec{u}$ and $\vec{v}$ is given by

$$<\vec{u},\vec{v}>= \left\|\vec{u}\right\|\cdot\left\|\vec{v}\right\|\cos(\vec{u},\vec{v}) = \cos(\beta - \alpha) = \cos(\alpha - \beta) = \cos\alpha\cos\beta + \sin\alpha\sin\beta$$

Thus,

$$\boxed{\cos(\alpha - \beta) = \cos\alpha\cos\beta + \sin\alpha\sin\beta,\ \forall\,\alpha,\beta \in R} \tag{1}$$

from which, we conclude that:

$$\boxed{\cos(\pi - \beta) = -\cos\beta\ \text{ şi }\ \cos(2\pi - \beta) = -\cos\beta,\ \forall\beta \in R}$$

Considering that $\beta \to -\beta$ and taking into account the imparity of the sine function and the parity of the cosine function one obtains

$$\boxed{\cos(\alpha + \beta) = \cos\alpha\cos\beta - \sin\alpha\sin\beta,\ \forall\,\alpha,\beta \in R} \tag{2}$$

For $\beta = \alpha$, (2) combined with (0) implies

$$\boxed{\cos 2\alpha = \cos^2\alpha - \sin^2\alpha = 2\cos^2\alpha - 1 = 1 - 2\sin^2\alpha,\ \forall\alpha \in R} \tag{3}$$

Thus,

$$\boxed{\cos\alpha = \pm\sqrt{\frac{1 + \cos 2\alpha}{2}},\ \sin\alpha = \pm\sqrt{\frac{1 - \cos 2\alpha}{2}},\ \forall\alpha \in R} \tag{4}$$

When $\alpha = \dfrac{\pi}{2}$, (1) implies

$$\cos\left(\frac{\pi}{2}-\beta\right)=\sin\beta, \text{ therefore } \sin\left(\frac{\pi}{2}-\beta\right)=\cos\beta, \; \forall\beta\in R \qquad (5)$$

Thus, from (1) and (5) it results

$$\sin(\alpha+\beta)=\cos\left(\frac{\pi}{2}-\alpha-\beta\right)=\cos\left(\frac{\pi}{2}-\alpha\right)\cos\beta+\sin\left(\frac{\pi}{2}-\alpha\right)\sin\beta=$$

$$=\sin\alpha\cos\beta+\sin\beta\cos\alpha, \; \forall\alpha,\beta\in R, \text{ which for } \beta\to-\beta \text{ generates} \qquad (6)$$

$$\sin(\alpha-\beta)=\sin\alpha\cos\beta-\sin\beta\cos\alpha, \; \forall\alpha,\beta\in R \qquad (7)$$

Hence,

$$\sin(\pi-\beta)=\sin\beta, \; \sin(2\pi-\beta)=-\sin\beta, \; \forall\beta\in R$$

(6) together with $\beta=\alpha$ leads to

$$\sin2\alpha=2\sin\alpha\cos\alpha, \; \forall\alpha\in R \qquad (8)$$

Considering that $\beta=2\alpha$ in (2) and using (3) together

with (0) we obtain:

$$\cos3\alpha=\cos(\alpha+2\alpha)=\cos\alpha\cos2\alpha-\sin\alpha\sin2\alpha=$$

$$=\cos\alpha(2\cos^2\alpha-1)-2\sin^2\alpha\cos\alpha=2\cos^3\alpha-\cos\alpha-2(1-\cos^2\alpha)\cos\alpha.$$

Consequently,

$$\cos3\alpha=4\cos^3\alpha-3\cos\alpha, \; \forall\alpha\in R \qquad (9)$$

Similarly, from (6) for $\beta=2\alpha$, (8), (3) and (0) we conclude

that

$$\sin3\alpha=\sin(\alpha+2\alpha)=\sin\alpha\cos2\alpha+\sin2\alpha\cos\alpha=$$

$$=\sin\alpha(1-2\sin^2\alpha)+2\sin\alpha(1-\sin^2\alpha), \; \forall\alpha\in R.$$

Thus,

$$\sin3\alpha=3\sin\alpha-4\sin^3\alpha, \; \forall\alpha\in R \qquad (10)$$

An interesting application of the formulae (2) and (6) is the
plane rotation of O center and α angle, which is attached to each

point $A(x,y)$ the point $A'(x',y')$ so that the angle $\widehat{AOA'}$ has α a radians (see Figure 34).

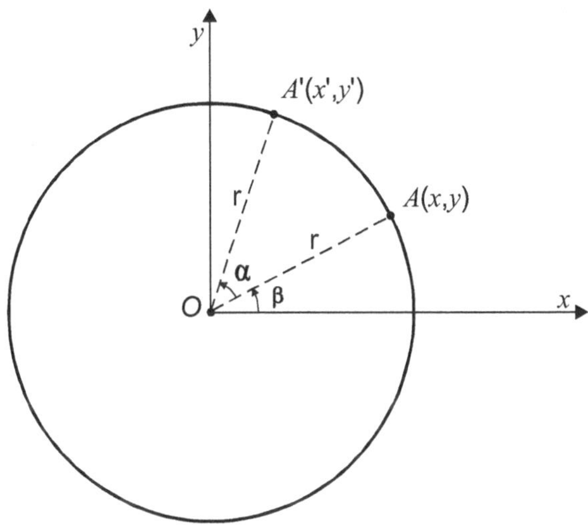

Figure 34: The rotation in any Euclidean plane

If the measure of the angle $(\overparen{Ox,OA})$ is equal to β radians

and $\|OA\| = \|OA'\| = r$, than

$$\begin{cases} x = r\cos\beta \\ y = r\sin\beta \end{cases} \quad \text{and} \quad \begin{cases} x' = r\cos(\alpha+\beta) \\ y' = r\sin(\alpha+\beta) \end{cases}.$$

So,

$$\begin{cases} x' = r(\cos\alpha\cos\beta - \sin\alpha\sin\beta) \\ y' = r(\sin\alpha\cos\beta + \sin\beta\cos\alpha) \end{cases} \Leftrightarrow \begin{cases} x' = x\cos\alpha - y\sin\alpha \\ y' = x\sin\alpha + y\cos\alpha \end{cases}$$

For the tangent function using (2) and (6) we obtain:

$$\text{tg}(\alpha+\beta) = \frac{\sin(\alpha+\beta)}{\cos(\alpha+\beta)} = \frac{\sin\alpha\cos\beta + \sin\beta\cos\alpha}{\cos\alpha\cos\beta - \sin\alpha\sin\beta},$$

which by simplifying with $\cos\alpha\cos\beta$, implies:

$$\boxed{\text{tg}(\alpha+\beta) = \frac{\text{tg}\,\alpha + \text{tg}\,\beta}{1 - \text{tg}\,\alpha\,\text{tg}\,\beta}, \quad \forall \alpha,\beta,\alpha+\beta \in R \setminus \{k\pi + \frac{\pi}{2} : k \in Z\}} \quad (11)$$

and

$$\text{tg}(\alpha - \beta) = \frac{\text{tg}\,\alpha - \text{tg}\,\beta}{1 + \text{tg}\,\alpha\,\text{tg}\,\beta}, \ \forall \alpha, \beta, \alpha - \beta \in R \setminus \left\{ k\pi + \frac{\pi}{2} : k \in Z \right\}, \qquad (12)$$

respectively, obtained from (11) by passing $\beta \to -\beta$ and taking into account the imparity of the tangent function.

From (5) one obtains that

$$\text{tg}\left(\frac{\pi}{2} - \beta\right) = \frac{\sin\left(\dfrac{\pi}{2} - \beta\right)}{\cos\left(\dfrac{\pi}{2} - \beta\right)} = \text{ctg}\,\beta, \ \forall \beta \in R \setminus \{k\pi : k \in Z\},$$

and (12) generates the relation

$$\text{tg}(\pi - \beta) = \text{tg}(2\pi - \beta) = -\text{tg}\,\beta, \ \forall \beta \in R \setminus \left\{ k\pi + \frac{\pi}{2} : k \in Z \right\}.$$

In the particular case, $\beta = \alpha$ (11) implies that

$$\text{tg}\,2\alpha = \frac{2\,\text{tg}\,\alpha}{1 - \text{tg}^2\,\alpha}, \qquad (13)$$

$$\forall \alpha \in R \setminus \left(\left\{ k\pi + \frac{\pi}{2} : k \in Z \right\} \cup \left\{ \frac{k\pi}{2} + \frac{\pi}{4} : k \in Z \right\} \right).$$

By succesively applying (11) it results that

$$\text{tg}(\alpha + \beta + \gamma) = \frac{\text{tg}\,\alpha + \text{tg}(\beta + \gamma)}{1 - \text{tg}\,\alpha\,\text{tg}(\beta + \gamma)} = \frac{\text{tg}\,\alpha + \dfrac{\text{tg}\,\beta + \text{tg}\,\gamma}{1 - \text{tg}\,\beta \cdot \text{tg}\,\gamma}}{1 - \text{tg}\,\alpha \cdot \dfrac{\text{tg}\,\beta + \text{tg}\,\gamma}{1 - \text{tg}\,\beta \cdot \text{tg}\,\gamma}}.$$

Thus,

$$\text{tg}(\alpha + \beta + \gamma) = \frac{\text{tg}\,\alpha + \text{tg}\,\beta + \text{tg}\,\gamma - \text{tg}\,\alpha \cdot \text{tg}\,\beta \cdot \text{tg}\,\gamma}{1 - \text{tg}\,\alpha \cdot \text{tg}\,\beta - \text{tg}\,\beta \cdot \text{tg}\,\gamma - \text{tg}\,\gamma \cdot \text{tg}\,\alpha} \qquad (14)$$

valid when all the expressions which intervene

have sense. So, if $\alpha = \beta = \gamma$, (14) induces

$$\boxed{\operatorname{tg}3\alpha = \frac{3\operatorname{tg}\alpha - \operatorname{tg}^3\alpha}{1-3\operatorname{tg}^2\alpha},}$$ (15)

$$\forall \alpha \in R \setminus (\{k\pi + \frac{\pi}{2} : k \in Z\} \cup \{\frac{k\pi}{3} + \frac{\pi}{6} : k \in Z\})$$

The definition of the contangent function together with

(2) and (6) leads to the relation

$$\operatorname{ctg}(\alpha+\beta) = \frac{\cos(\alpha+\beta)}{\sin(\alpha+\beta)} = \frac{\cos\alpha\cos\beta - \sin\alpha\sin\beta}{\sin\alpha\cos\beta + \sin\beta\cos\alpha},$$

which generates, by simplifying with $\sin\alpha\sin\beta$,

$$\boxed{\operatorname{ctg}(\alpha+\beta) = \frac{\operatorname{ctg}\alpha\operatorname{ctg}\beta - 1}{\operatorname{ctg}\alpha + \operatorname{ctg}\beta}, \ \forall \alpha, \beta, \alpha+\beta \in R \setminus \{k\pi : k \in Z\}}$$ (16)

which for $\beta \to -\beta$ together with the imparity of the

contangent function implies

$$\boxed{\operatorname{ctg}(\alpha-\beta) = \frac{\operatorname{ctg}\alpha\operatorname{ctg}\beta + 1}{\operatorname{ctg}\beta - \operatorname{ctg}\alpha}, \ \forall \alpha, \beta, \alpha-\beta \in R \setminus \{k\pi : k \in Z\}}$$ (17)

According to (1), (5) and (7),

$$\operatorname{ctg}\left(\frac{\pi}{2} - \alpha\right) = \operatorname{tg}\alpha, \ \forall \alpha \in R \setminus \{k\pi + \frac{\pi}{2} : k \in Z\},$$

$$\operatorname{ctg}(\pi - \alpha) = -\operatorname{ctg}\alpha = \operatorname{ctg}(2\pi - \alpha), \ \forall \alpha \in R \setminus \{k\pi : k \in Z\}$$

If $\beta = \alpha$, from (16) we obtain

$$\boxed{\operatorname{ctg}2\alpha = \frac{\operatorname{ctg}^2\alpha - 1}{2\operatorname{ctg}\alpha}, \ \forall \alpha \in R \setminus (\{k\pi : k \in Z\} \cup \{\frac{k\pi}{2} : k \in Z\})}$$ (18)

and succesive applications of (16) generate

$$\operatorname{ctg}(\alpha+\beta+\gamma) = \frac{\operatorname{ctg}\alpha \cdot \operatorname{ctg}(\beta+\gamma) - 1}{\operatorname{ctg}\alpha + \operatorname{ctg}(\beta+\gamma)} = \frac{\operatorname{ctg}\alpha \cdot \dfrac{\operatorname{ctg}\beta \cdot \operatorname{ctg}\gamma - 1}{\operatorname{ctg}\beta + \operatorname{ctg}\gamma} - 1}{\operatorname{ctg}\alpha + \dfrac{\operatorname{ctg}\beta \cdot \operatorname{ctg}\gamma - 1}{\operatorname{ctg}\beta + \operatorname{ctg}\gamma}}.$$

Thus,

$$\operatorname{ctg}(\alpha+\beta+\gamma)=\frac{\operatorname{ctg}\alpha\cdot\operatorname{ctg}\beta\cdot\operatorname{ctg}\gamma-\operatorname{ctg}\alpha-\operatorname{ctg}\beta-\operatorname{ctg}\gamma}{\operatorname{ctg}\alpha\cdot\operatorname{ctg}\beta+\operatorname{ctg}\beta\cdot\operatorname{ctg}\gamma+\operatorname{ctg}\gamma\cdot\operatorname{ctg}\alpha-1}$$

(19)

each time it has sense.

If $\alpha=\beta=\gamma$, (19) implies that

$$\operatorname{ctg}3\alpha=\frac{\operatorname{ctg}^{3}\alpha-3\operatorname{ctg}\alpha}{3\operatorname{ctg}^{2}\alpha-1}$$

(20)

$$\forall\alpha\in R\setminus(\{k\pi:k\in Z\}\cup\{\frac{k\pi}{3}:k\in Z\})$$

(1), (2), (6) and (7) justifies

(21)

$$\cos\alpha\cos\beta=\frac{\cos(\alpha+\beta)+\cos(\alpha-\beta)}{2},$$

$$\sin\alpha\sin\beta=\frac{\cos(\alpha-\beta)-\cos(\alpha+\beta)}{2},$$

$$\sin\alpha\cos\beta=\frac{\sin(\alpha+\beta)+\sin(\alpha-\beta)}{2},\ \forall\alpha,\beta\in R$$

and

$$\cos\alpha+\cos\beta=\cos\left(\frac{\alpha+\beta}{2}+\frac{\alpha-\beta}{2}\right)+\cos\left(\frac{\alpha+\beta}{2}-\frac{\alpha-\beta}{2}\right)=$$

(22)

$$=2\cos\frac{\alpha+\beta}{2}\cos\frac{\alpha-\beta}{2},$$

$$\cos\alpha-\cos\beta=-2\sin\frac{\alpha+\beta}{2}\sin\frac{\alpha-\beta}{2},$$

$$\sin\alpha+\sin\beta=\sin\left(\frac{\alpha+\beta}{2}+\frac{\alpha-\beta}{2}\right)+\sin\left(\frac{\alpha+\beta}{2}-\frac{\alpha-\beta}{2}\right)=$$

$$=2\sin\frac{\alpha+\beta}{2}\cos\frac{\alpha-\beta}{2},$$

$$\sin\alpha-\sin\beta=2\sin\frac{\alpha-\beta}{2}\cos\frac{\alpha+\beta}{2},\ \forall\alpha,\beta\in R$$

The following obvious relations obtained from (0), (3) and (8) by

$$\text{replacing} \quad \alpha \to \frac{\alpha}{2}: \quad \sin\alpha = \frac{2\sin\frac{\alpha}{2}\cos\frac{\alpha}{2}}{\cos^2\frac{\alpha}{2}+\sin^2\frac{\alpha}{2}}, \quad \cos\alpha = \frac{\cos^2\frac{\alpha}{2}-\sin^2\frac{\alpha}{2}}{\cos^2\frac{\alpha}{2}+\sin^2\frac{\alpha}{2}},$$

after simplifying with $\cos^2\frac{\alpha}{2}$, they lead to:

$$\boxed{\sin\alpha = \frac{2\,\text{tg}\frac{\alpha}{2}}{1+\text{tg}^2\frac{\alpha}{2}}, \quad \cos\alpha = \frac{1-\text{tg}^2\frac{\alpha}{2}}{1+\text{tg}^2\frac{\alpha}{2}}, \quad \forall\alpha\in R\setminus\{(2k+1)\pi:k\in Z\}} \qquad (23)$$

and

$$\boxed{\text{tg}\,\alpha = \frac{2\,\text{tg}\frac{\alpha}{2}}{1-\text{tg}^2\frac{\alpha}{2}}, \quad \forall\alpha\in R\setminus\left(\{k\pi+\frac{\pi}{2}:k\in Z\}\cup\{(2k+1)\pi:k\in Z\}\right)} \qquad (24)$$

$$\boxed{\text{ctg}\,\alpha = \frac{1-\text{tg}^2\frac{\alpha}{2}}{2\,\text{tg}\frac{\alpha}{2}}, \quad \forall\alpha\in R\setminus\{k\pi:k\in Z\}}$$

5.2. IDENTITIES

1. $\quad\boxed{\sin\alpha+\sin 2\alpha+...+\sin n\alpha = \frac{\sin\frac{n\alpha}{2}\sin\frac{n+1}{2}\alpha}{\sin\frac{\alpha}{2}},}$

$\boxed{\forall\alpha\in R\setminus\{2k\pi:k\in Z\}}$

Indeed:

$$\sum_{k=1}^{n}\sin k\alpha = \frac{1}{2\sin\dfrac{\alpha}{2}}\sum_{k=1}^{n}2\sin\frac{\alpha}{2}\sin k\alpha = \frac{1}{2\sin\dfrac{\alpha}{2}}\sum_{k=1}^{n}(\cos\frac{2k-1}{2}\alpha - \cos\frac{2k+1}{2}\alpha) =$$

$$= \frac{\sin\dfrac{n\alpha}{2}\sin\dfrac{n+1}{2}\alpha}{\sin\dfrac{\alpha}{2}}.$$

2. By using the relation:

$$\sin[\alpha+(k-1)\beta] = \left[\cos\left(\alpha+\frac{2k-3}{2}\beta\right) - \cos\left(\alpha+\frac{2k-1}{2}\beta\right)\right]\bigg/ 2\sin\frac{\beta}{2}$$

for $\forall\alpha\in R,\ \beta\in R\setminus\{2p\pi : p\in Z\}$ and $k=\overline{1,n}$ we obtain

$$\sum_{k=1}^{n}\sin[\alpha+(k-1)\beta] = \frac{\sin\left(\alpha+\dfrac{n-1}{2}\beta\right)\sin\dfrac{n\beta}{2}}{\sin\dfrac{\beta}{2}},$$

$$\forall\alpha\in R,\ \beta\in R\setminus\{2p\pi : p\in Z\}$$

3. Similarly it is shown that:

$$\sum_{k=1}^{n}\cos[\alpha+(k-1)\beta] = \frac{\cos\left(\alpha+\dfrac{n-1}{2}\beta\right)\sin\dfrac{n\beta}{2}}{\sin\dfrac{\beta}{2}},$$

$$\forall\alpha\in R,\ \beta\in R\setminus\{2p\pi : p\in Z\}$$

4. Because $1+\sec 2^{k}\alpha = \dfrac{\operatorname{tg}2^{k}\alpha}{\operatorname{tg}2^{k-1}\alpha}$, $n\in N^{*}$, $k=\overline{1,n}$, it results that

$$(1+\sec 2\alpha)(1+\sec 4\alpha)\cdot\ldots\cdot(1+\sec 2^{n}\alpha) = \frac{\operatorname{tg}2^{n}\alpha}{\operatorname{tg}\alpha}$$

under the adequate conditions of existence.

5. From the equality $\sin\dfrac{\alpha}{2^k}\cdot\cos\dfrac{\alpha}{2^k}=\dfrac{1}{2}\sin\dfrac{\alpha}{2^{k-1}}$,

$\forall\alpha\in R,\ n\in N^*,\ k=\overline{1,n}$ we can deduce, immediately that

$$\boxed{\cos\dfrac{\alpha}{2}\cdot\cos\dfrac{\alpha}{2^2}\cdot\ldots\cdot\cos\dfrac{\alpha}{2^n}=\dfrac{\sin\alpha}{2^n\sin\dfrac{\alpha}{2^n}},\quad \forall n\in N^*,\ \alpha\neq 2^n\,p\pi,\ p\in Z}$$

6. By considering successively that $\beta=\dfrac{\alpha}{2^k}$, $k=\overline{1,n}$, $n\in N^*$ in the next obvious relation

$\operatorname{ctg}\beta-\operatorname{tg}\beta=2\operatorname{ctg}2\beta$ one obtains $\dfrac{1}{2^k}\operatorname{ctg}\dfrac{\alpha}{2^k}-\dfrac{1}{2^{k-1}}\operatorname{ctg}\dfrac{\alpha}{2^{k-1}}=\dfrac{1}{2^k}\operatorname{tg}\dfrac{\alpha}{2^k}$,

$k=\overline{1,n}$, thus

$$\boxed{\sum_{k=1}^{n}\dfrac{1}{2^k}\operatorname{tg}\dfrac{\alpha}{2^k}=\dfrac{1}{2^n}\operatorname{ctg}\dfrac{\alpha}{2^n}-\operatorname{ctg}\alpha}$$

7. From the relation

$$\operatorname{ctg}2^{k-1}\alpha-\operatorname{ctg}2^k\alpha=\dfrac{1}{\sin 2^k\alpha},\quad \forall n\in N^*,\ k=\overline{1,n},\ \alpha\neq p\pi,\ \dfrac{p\pi}{2^k},\ p\in Z$$

it follows that

$$\boxed{\sum_{k=1}^{n}\dfrac{1}{\sin 2^k\alpha}=\operatorname{ctg}\alpha-\operatorname{ctg}2^n\alpha}$$

8. The property $\operatorname{arctg}x-\operatorname{arctg}y=\operatorname{arctg}\dfrac{x-y}{1+xy},\quad \forall x,y,\ x\cdot y\neq-1$

implies that

$$\boxed{\sum_{k=1}^{n}\operatorname{arctg}\dfrac{1}{1+k+k^2}=\sum_{k=1}^{n}\left(\operatorname{arctg}\dfrac{1}{k}-\operatorname{arctg}\dfrac{1}{k+1}\right)=\dfrac{\pi}{4}-\operatorname{arctg}\dfrac{1}{n+1},\quad \forall n\in N^*}$$

9. $\arcsin x,\ \dfrac{\pi}{2}-\arccos x\in\left[-\dfrac{\pi}{2},\dfrac{\pi}{2}\right],\quad \forall x\in[-1,1]$

and   as $\sin(\arcsin x) = \sin\left(\dfrac{\pi}{2} - \arccos x\right) = x, \ \forall x \in [-1,1]$ the injectivity of the restriction of the sine function at the interval $\left[-\dfrac{\pi}{2}, \dfrac{\pi}{2}\right]$ leads to the relation

$$\arcsin x + \arccos x = \frac{\pi}{2}, \ \forall x \in [-1,1]$$

Similarly, it is proven that

$$\operatorname{arctg} y + \operatorname{arcctg} y = \frac{\pi}{2}, \ \forall y \in R$$

10.   (14) justifies the following property:

$$\operatorname{arctg} x + \operatorname{arctg} y + \operatorname{arctg} z = \operatorname{arctg} \frac{x + y + z - xyz}{1 - xy - yz - xz}, \ \forall x, y, z \in R$$

$$\text{with } xy + yz + xz \neq 1.$$

5.3. INEQUALITIES

1. The obvious inequality

$$\left(x \sin \alpha + y \cos \alpha\right)^2 \leq \left(x^2 + y^2\right)\left(\sin^2 \alpha + \cos^2 \alpha\right), \qquad \forall \alpha, x, y \in R,$$

together with (0) leads to

$$\left|x \sin \alpha + y \cos \alpha\right| \leq \sqrt{x^2 + y^2}, \ \forall \alpha, x, y \in R$$

with equality only for $x \cos \alpha = y \sin \alpha$.

Particularly, $\left|\dfrac{1 \pm \sin 2\alpha}{\sin \alpha \pm \cos \alpha}\right| \leq \sqrt{2}, \ \forall \alpha \in R \setminus \left\{k\pi \pm \dfrac{\pi}{4} : k \in Z\right\}$ where the signes correspond to each other.

2. The strictly concavity of the sine function on $[0, \pi]$ and of cosine function on $\left[0, \dfrac{\pi}{2}\right]$ leads to:

$$\sin\alpha+\sin\beta+\sin\gamma\leq 3\sin\frac{\alpha+\beta+\gamma}{3}=\frac{3\sqrt{3}}{2}, \qquad \forall\alpha,\beta,\gamma\in(o,\pi) \qquad \text{cu}$$

$$\boxed{\alpha+\beta+\gamma=\pi}\ \text{and}$$

$$\cos\alpha+\cos\beta+\cos\gamma\leq 3\cos\frac{\alpha+\beta+\gamma}{3}=\frac{3\sqrt{3}}{2},$$

$$\forall\alpha,\beta,\gamma\in\left[0,\frac{\pi}{2}\right],\ \alpha+\beta+\gamma=\frac{\pi}{2},$$

respectively

3. The usual inequality between the arithmetical and the geometrical means combined with the strictly concavity of the sine function on $(0,\pi)$ implies that

$$\sin\frac{\alpha}{2}\cdot\sin\frac{\beta}{2}\cdot\sin\frac{\gamma}{2}\leq\left(\frac{\sin\frac{\alpha}{2}+\sin\frac{\beta}{2}+\sin\frac{\gamma}{2}}{3}\right)^{3}\leq\left(\sin\frac{\alpha+\beta+\gamma}{6}\right)^{3}=\frac{1}{8},$$

$\forall\alpha,\beta,\gamma\in(o,\pi)$ cu $\alpha+\beta+\gamma=\pi$, with equality only if $\alpha=\beta=\gamma=\dfrac{\pi}{3}$.

4. By analogy, using the strictly concavity of the

cosine function on the interval $\left[0,\dfrac{\pi}{2}\right]$ one obtains

$$\cos\frac{\alpha}{2}\cdot\cos\frac{\beta}{2}\cdot\cos\frac{\gamma}{2}\leq\left(\frac{\cos\frac{\alpha}{2}+\cos\frac{\beta}{2}+\cos\frac{\gamma}{2}}{3}\right)^{3}\leq\left(\cos\frac{\alpha+\beta+\gamma}{6}\right)^{3}=\frac{3\sqrt{3}}{8}$$

$\forall\alpha,\beta,\gamma\in(o,\pi)$ cu $\alpha+\beta+\gamma=\pi$, equlity occuring if $\alpha=\beta=\gamma=\dfrac{\pi}{3}$.

5. $\boxed{\sin\alpha_1\cos\alpha_2\cos\alpha_3...\cos\alpha_n+\cos\alpha_1\sin\alpha_2\sin\alpha_3...\sin\alpha_n\leq 1}$,

$$\forall n\in N^{*},\ n\geq 2,\ \alpha_i\in R,\ i=\overline{1,n}$$

Indeed,

$$\left|\sin\alpha_1\cos\alpha_2...\cos\alpha_n+\cos\alpha_1\sin\alpha_2...\sin\alpha_n\right|\leq\left|\sin\alpha_1\cos\alpha_2\right|+\left|\cos\alpha_1\sin\alpha_2\right|=$$

$$=\pm\sin(\alpha_1\pm\alpha_2)\leq1.$$

6. For each $n\in N^*$, $n\geq2$ the function $x\to x^n$ is strictly convex

on $[0,+\infty)$, thus $\left(\dfrac{x_1+x_2}{2}\right)^n\leq\dfrac{1}{2}\left(x_1^n+x_2^n\right),\ \forall x_1x_2\in[0,+\infty).$

Therefore,

$$\frac{1}{2^n}=\left(\frac{\sin^2\alpha+\cos^2\alpha}{2}\right)^n\leq\frac{1}{2}\left(\sin^{2n}\alpha+\cos^{2n}\alpha\right)\leq\frac{1}{2}\left(\sin^2\alpha+\cos^2\alpha\right)^n=\frac{1}{2},$$

that is

$$\boxed{\ \frac{1}{2^{n-1}}\leq\sin^{2n}\alpha+\cos^{2n}\alpha\leq1,\ \forall\alpha\in R\ }$$

7. Because the restriction of the tangent function at

$\left(0,\dfrac{\pi}{2}\right)$ is strictly increasing it follows that

$$\operatorname{tg}\alpha_1\cdot\cos\alpha_i\leq\sin\alpha_i\leq\operatorname{tg}\alpha_n\cdot\cos\alpha_i,\ \forall n\in N^*,\ \alpha_i\in\left(0,\frac{\pi}{2}\right)\ \text{cu}$$

$\alpha_1\leq\alpha_i\leq\alpha_n,\ i=\overline{1,n}.$

So,

$$\boxed{\ \operatorname{tg}\alpha_1\leq\frac{\displaystyle\sum_{i=1}^{n}\sin\alpha_i}{\displaystyle\sum_{i=1}^{n}\cos\alpha_i}\leq\operatorname{tg}\alpha_n\ }$$

with equality iff $\alpha_1=\alpha_2=...=\alpha_n.$

8. The strictly convexity of the tangent function on

$\left[0,\dfrac{\pi}{2}\right)$ implies that

$$\boxed{\operatorname{tg}\frac{\alpha}{2}+\operatorname{tg}\frac{\beta}{2}+\operatorname{tg}\frac{\gamma}{2}\geq 3\operatorname{tg}\left(\frac{\alpha+\beta+\gamma}{6}\right)=\sqrt{3},\qquad \forall\,\alpha,\beta,\gamma\in(o,\pi)}$$ cu

$$\boxed{\alpha+\beta+\gamma=\pi}\,.$$

9. The customary inequality between the arithmetical and the harmonical means implies that

$$\boxed{\left(\sum_{i=1}^{n}\operatorname{tg}\alpha_i\right)\left(\sum_{i=1}^{n}\operatorname{ctg}\alpha_i\right)\geq n^2,\ \ \forall n\in N^*,\ \alpha_i\in\left(0,\frac{\pi}{2}\right),\ i=\overline{1,n}}$$

with equality if only if $\alpha_1=\alpha_2=...=\alpha_n$.

10. The strictly convexity of the cotangent function on $\left(0,\dfrac{\pi}{2}\right]$ induces:

$$\boxed{\operatorname{ctg}\left(\frac{\displaystyle\sum_{i=1}^{n}\alpha_i}{n}\right)\leq\frac{1}{n}\sum_{i=1}^{n}\operatorname{ctg}\alpha_i,\ \forall n\in N^*,\ \alpha_i\in\left(0,\frac{\pi}{2}\right),\ i=\overline{1,n}}\,,$$

and the strictly concavity of the same function on $\left[\dfrac{\pi}{2},\pi\right)$ leads to

$$\boxed{\operatorname{ctg}\left(\frac{\displaystyle\sum_{i=1}^{n}\beta_i}{n}\right)\geq\frac{1}{n}\sum_{i=1}^{n}\operatorname{ctg}\beta_i,\ \forall n\in N^*,\ \beta_i\in\left[\frac{\pi}{2},\pi\right),\ i=\overline{1,n}}\,.$$

5.4. EQUATIONS

1. $\sec x+\operatorname{cosec} x=\left(\sin x+\cos x\right)\left(\sec^2 x+\operatorname{cosec}^2 x\right),$

$$x\in R\setminus\left(\{k\pi:k\in Z\}\cup\left\{2k\pi\pm\frac{\pi}{2}:k\in Z\right\}\right)$$

The equation is also written:

$$\frac{\sin x + \cos x}{\sin x \cos x} = \frac{\sin x + \cos x}{\sin^2 x \cos^2 x} \Leftrightarrow \sin x + \cos x = 0 \text{ sau } \sin x \cos x = 1 \Leftrightarrow$$

$$\sin\left(x + \frac{\pi}{4}\right) = 0 \text{ sau } \sin 2x = 2 \Leftrightarrow x = k\pi - \frac{\pi}{4}, \ k \in Z.$$

2. $\sin^{2n} x + \cos^{2n} x = 1, \ n \in N.$

For $n = 0$ the equation has no solution, and if $n = 1$ the

set of the solutions is R. Let $n \geq 2$ be any natural

number. Then,

$$1 = \left(\sin^2 x + \cos^2 x\right)^n = \sin^{2n} x + \cos^{2n} x + n\sin^{2n-2} x \cos^2 x + \ldots + n\sin^2 x \cos^{2n-2} x$$

. So, the equation is equivalent to $\sin x = 0$ or $\cos x = 0$ $\Leftrightarrow$

$$x \in \{k\pi : k \in Z\} \cup \left\{2k\pi \pm \frac{\pi}{2} : k \in Z\right\}$$

3. $\arcsin\left(\sin x\right) = t \xleftarrow{\ t \in \left[-\frac{\pi}{2}, \frac{\pi}{2}\right]\ } \sin x = \sin t \Leftrightarrow x = k\pi + \left(-1\right)^k t, \ k \in Z.$

$\arccos\left(\cos x\right) = t \xleftarrow{\ t \in [0, \pi]\ } \cos x = \cos t \Leftrightarrow x = 2k\pi \pm t, \ k \in Z.$

$\operatorname{arctg}\left(\operatorname{tg} x\right) = t \xleftarrow{\ t \in \left(-\frac{\pi}{2}, \frac{\pi}{2}\right)\ } \operatorname{tg} x = \operatorname{tg} t \Leftrightarrow x = k\pi + t, \ k \in Z.$

$\operatorname{arcctg}\left(\operatorname{ctg} x\right) = t \xleftarrow{\ t \in (0, \pi)\ } \operatorname{ctg} x = \operatorname{ctg} t \Leftrightarrow x = k\pi + t, \ k \in Z.$

4. $\left(1 - \cos x\right)^2 + \cos^2\left(2 - x\right) = 0$ has no solution in R because

it is equivalent to the sistem

$$\begin{cases} \cos x = 1 \\ \cos(2 - x) = 0 \end{cases} \Leftrightarrow \begin{cases} x = 2k\pi, \ k \in Z \\ 2 - x = 2q\pi \pm \dfrac{\pi}{2}, q \in Z \end{cases} \Rightarrow 4 = 4k\pi + 4q\pi \pm \pi,$$

contradicting $\pi \in R \setminus Q.$

5. The equation $2\cos\big(|x|-1\big)=|x|+\dfrac{1}{|x|}$, $x\in R^{*}$ is equivalent with

$|x|=1$, because $2\cos\big(|x|-1\big)\le 2\le |x|+\dfrac{1}{|x|}$, $\forall x\in R^{*}$, thus the equation

has the solutions $x=\pm 1$.

6. Each of the equations $\big[\sin x\big]=\alpha$ and $\big[\cos x\big]=\beta$ ...has no

solution for $\alpha,\beta\notin\{-1,0,1\}$ because $\big[\sin x\big],\big[\cos x\big]\in\{-1,0,1\}$.

7. The identity $\arcsin u+\arccos u=\dfrac{\pi}{2}$, $\forall u\in[-1,1]$ implies the

equivalence of the equation $\arcsin 2x+\arccos(1-2x)=\dfrac{\pi}{2}$ with the

equation $2x=1-2x\Leftrightarrow x=\dfrac{1}{4}$, and $\operatorname{arctg} y+\operatorname{arcctg} y=\dfrac{\pi}{2}$, $\forall y\in R$

implies $\operatorname{arctg}\sqrt{x}+\operatorname{arcctg} x=\dfrac{\pi}{2}\Leftrightarrow \sqrt{x}=x$, thus $x=0$.

8. $\operatorname{tg} x\cdot\operatorname{tg} 2x+\operatorname{tg} 2x\cdot\operatorname{tg} 3x+\operatorname{tg} 3x\cdot\operatorname{tg} x=1$,

$$x\in R\setminus\left(\left\{k\pi+\dfrac{\pi}{2}:k\in Z\right\}\cup\left\{\dfrac{k\pi}{2}+\dfrac{\pi}{4}:k\in Z\right\}\cup\left\{\dfrac{k\pi}{3}+\dfrac{\pi}{6}:k\in Z\right\}\right)$$

In accordance with (14) in the section 5.1, x is solution to the given equation if and only if there doesn't exist $\operatorname{tg}(x+2x+3x)=\operatorname{tg} 6x$, that is, iff $x\in\left\{\dfrac{k\pi}{6}+\dfrac{\pi}{12}:k\in Z\right\}$.

9. According to the same mentioned relation in the previous example, the equation

$$\operatorname{tg} x+\operatorname{tg} 2x+\operatorname{tg} 3x=\operatorname{tg} x\cdot\operatorname{tg} 2x\cdot\operatorname{tg} 3x \quad\text{examined on the set}$$

$$R\setminus\left(\left\{k\pi+\dfrac{\pi}{2}:k\in Z\right\}\cup\left\{\dfrac{k\pi}{2}+\dfrac{\pi}{4}:k\in Z\right\}\cup\left\{\dfrac{k\pi}{3}+\dfrac{\pi}{6}:k\in Z\right\}\right)$$

is equivalent to $\operatorname{tg}6x = 0 \Leftrightarrow x = \dfrac{k\pi}{6}, \ k \in Z.$

10. $\operatorname{arcctg}x - \operatorname{arcctg}(x+1) = \dfrac{\pi}{4}, \ x \in R \Leftrightarrow \operatorname{arcctg}\left[x(x+1)+1\right] = \dfrac{\pi}{4}$

$$\Leftrightarrow \quad x(x+1) = 0 \Leftrightarrow x \in \{-1,0\}.$$

5.5. SYSTEMS OF EQUATIONS

1. The equation in

$x, y, z \in R \quad \sin(x+y)\cdot\sin(y+z)\cdot\sin(z+x) = 1$ is equivalent to the following sequence of systems:

$$\begin{cases} \sin(x+y) = 1 \\ \sin(y+z) = 1 \\ \sin(z+x) = 1 \end{cases} \text{ or } \begin{cases} \sin(x+y) = -1 \\ \sin(y+z) = -1 \\ \sin(z+x) = 1 \end{cases} \text{ or } \begin{cases} \sin(x+y) = -1 \\ \sin(y+z) = 1 \\ \sin(z+x) = -1 \end{cases} \text{ or }$$

$$\begin{cases} \sin(y+z) = -1 \\ \sin(x+y) = 1 \\ \sin(z+x) = -1 \end{cases} \quad \text{which can be solved by using the basic}$$

equations $\sin u = -1 \Leftrightarrow$

$$u = q\pi + (-1)^{q+1}\cdot\dfrac{\pi}{2}, \ q \in Z$$

respectively $\sin v = 1 \Leftrightarrow v = k\pi + (-1)^{k}\cdot\dfrac{\pi}{2}, \ k \in Z.$

2. The system $\begin{cases} \sin x + \sin y = a \\ \cos x + \cos y = b \end{cases}$ doesn't have solutions for

any $a, b \in R$ with $a^2 + b^2 > 4.$

Indeed, supposing the contrary, by square raising both relations and addition, we obtain:

$$\cos(x-y) = \dfrac{a^2 + b^2 - 2}{2} > 1, \text{ a contradiction.}$$

3. $\begin{cases} \operatorname{tg} x + \operatorname{tg} y = 3 \\ \operatorname{ctg} x + \operatorname{ctg} y = 1 \end{cases}$ $\Leftrightarrow$ $\begin{cases} \operatorname{tg} x + \operatorname{tg} y = 3 \\ \operatorname{tg} x \cdot \operatorname{tg} y = 3 \end{cases}$ $\Leftrightarrow$

$\begin{cases} \operatorname{tg}^2 x - 3 \operatorname{tg} x + 3 = 0 \\ \operatorname{tg} x \cdot \operatorname{tg} y = 3 \end{cases}$ first   equation   without   solution

because $\operatorname{tg}^2 x - 3\operatorname{tg} x + 3 = \left(\operatorname{tg} x - \dfrac{3}{2}\right)^2 + \dfrac{3}{4} > 0,$

$\forall x \in R \setminus \left\{ k\pi + \dfrac{\pi}{2} : k \in Z \right\}.$ In   consequence,   the   system

doesn't have solutions.

4. $\begin{cases} \operatorname{tg} x + \operatorname{tg} y + \operatorname{tg} z = \operatorname{tg} x \cdot \operatorname{tg} y \cdot \operatorname{tg} z \\ x - y + z = 2\pi \\ x + y - z = 3\pi \end{cases}$ $\Leftrightarrow$ $\begin{cases} x + y + z = k\pi, \ k \in Z \\ x - y + z = 2\pi \\ x + y - z = 3\pi \end{cases}$ $\Leftrightarrow$

$\begin{cases} x = \dfrac{5\pi}{2}, \\ y = \dfrac{k-2}{2}\pi, \\ z = \dfrac{k-3}{2}\pi, \ k \in Z \end{cases}$

5. $\begin{cases} \arcsin x + \arcsin y = 0 \\ x^2 + y^2 = 1 \end{cases}$ $\Leftrightarrow$ $\begin{cases} x = -\dfrac{\sqrt{2}}{2} \\ y = \dfrac{\sqrt{2}}{2} \end{cases}$ or $\begin{cases} x = \dfrac{\sqrt{2}}{2} \\ y = -\dfrac{\sqrt{2}}{2} \end{cases}.$

6. $\begin{cases} \arccos x^2 + \arccos\left(y^2 - 4\right) = \pi \\ x^2 - y^2 = 4 \end{cases}$ $\Leftrightarrow$ $\begin{cases} x^2 + y^2 = 4 \\ x^2 - y^2 = 4 \end{cases}$ $\Leftrightarrow$

$\begin{cases} x = 2 \\ y = 0 \end{cases}$ or $\begin{cases} x = -2 \\ y = 0 \end{cases}$

both solutions being unacceptable because $|x| \le 1.$

7. $\begin{cases} \arctg x - \arctg y = \dfrac{\pi}{4} \\ xy = -\dfrac{1}{2} \end{cases} \Leftrightarrow \begin{cases} \dfrac{x-y}{1+xy} = 1 \\ xy = -\dfrac{1}{2} \end{cases}$

which doesn't have real solutions.

8. $\begin{cases} \sin^2 x - \sin^2(y-z) = 1 \\ \sin^2 y - \sin^2(z-x) = 1 \\ \sin^2 z - \sin^2(x-y) = 1 \end{cases} \Leftrightarrow \begin{cases} \cos 2(y-z) - \cos 2x = 2 \\ \cos 2(z-x) - \cos 2y = 2 \\ \cos 2(x-y) - \cos 2z = 2 \end{cases} \Leftrightarrow$

$\begin{cases} \sin(x+y-z)\cdot\sin(x-y+z) = 1 \\ \sin(y+z-x)\cdot\sin(x+y-z) = 1 \\ \sin(x-y+z)\cdot\sin(y+z-x) = 1 \end{cases} \Leftrightarrow$

$\begin{cases} \sin(x+y-z) = \pm 1 \\ \sin(y+z-x) = \pm 1 \\ \sin(x+z-y) = \pm 1 \end{cases} \Leftrightarrow$

$\Leftrightarrow \begin{cases} x+y-z = k\pi \pm (-1)^k \dfrac{\pi}{2}, \\ y+z-x = q\pi + (-1)^q \dfrac{\pi}{2}, \\ x+z-y = h\pi + (-1)^h \dfrac{\pi}{2}, \ h,k,q \in Z \end{cases}$

9. The system

$\begin{cases} \arcctg(x+y) + \arcctg(x-y) = \dfrac{\pi}{4} \\ x^2 + y^2 = 2x + 3 \end{cases}$

considered in the set

$$\left\{ (x,y) \in R^2 : x^2 + y^2 > 3 \right\}$$

is equivalent to

$$\begin{cases} arc\,ctg\,\dfrac{x^2-y^2-1}{2x}=\dfrac{\pi}{4} \\ x^2+y^2-3=2x \end{cases} \Leftrightarrow$$

$$\begin{cases} x^2-y^2-1=x^2+y^2-3 \\ x^2+y^2-2x-3 \end{cases} \Leftrightarrow$$

$$\begin{cases} y=\pm 1 \\ x=1\pm\sqrt{3} \end{cases}.$$

10. $\begin{cases} tg\,x+ctg\,y=1 \\ tg\,y+ctg\,x=1 \end{cases} \Leftrightarrow \begin{cases} tg\,x=tg\,y \\ tg\,y+\dfrac{1}{tg\,y}=1 \end{cases} \Leftrightarrow \begin{cases} x=k\pi+y,\ k\in Z \\ tg^2\,y-tg\,y+1=0 \end{cases}$

the last equation without solution because

$$tg^2\,y-tg\,y+1=\left(tg\,y-\dfrac{1}{2}\right)^2+\dfrac{3}{4}>0,\ \forall y\in R\setminus\left\{k\pi+\dfrac{\pi}{2}:k\in Z\right\}.$$

5.6. INEQUATIONS

The monotony together with the announced periodicity and the sign of the elementary trigonometric functions generates the following solutions for the inequations indicated below.

1. $\sin x+\cos x>\sqrt{\dfrac{3}{2}} \Leftrightarrow \sqrt{2}\sin\left(x+\dfrac{\pi}{4}\right)>\sqrt{\dfrac{3}{2}} \Leftrightarrow \sin\left(x+\dfrac{\pi}{4}\right)>\sin\dfrac{\pi}{3}$

$\Leftrightarrow \qquad 2k\pi+\dfrac{\pi}{3}<x+\dfrac{\pi}{4}<2k\pi+\dfrac{2\pi}{3},\ k\in Z \qquad \Leftrightarrow$

$$x\in\bigcup_{k\in Z}\left(2k\pi+\dfrac{\pi}{12},\ 2k\pi+\dfrac{5\pi}{12}\right).$$

2. $\cos x-\sin x<\dfrac{1}{\sqrt{2}} \Leftrightarrow \cos\left(x-\dfrac{\pi}{4}\right)<\cos\dfrac{\pi}{3} \Leftrightarrow$

$$2k\pi+\dfrac{\pi}{3}<x-\dfrac{\pi}{4}<2k\pi+\dfrac{5\pi}{3},\ k\in Z$$

$$\Leftrightarrow \ x \in \bigcup_{k \in Z}\left(2k\pi + \frac{7\pi}{12}, \ (2k+2)\pi - \frac{\pi}{12}\right).$$

3. $\dfrac{\operatorname{tg} x - 1}{\operatorname{tg} x + 1} > \sqrt{3}, \ x \in R \setminus \left(\left\{k\pi + \dfrac{\pi}{2} : k \in Z\right\} \cup \left\{k\pi - \dfrac{\pi}{4} : k \in Z\right\}\right)$

is equivalent to

$$\operatorname{tg}\left(x - \frac{\pi}{4}\right) > \operatorname{tg}\frac{\pi}{3} \ \Leftrightarrow \ k\pi + \frac{\pi}{3} < x - \frac{\pi}{4} < k\pi + \frac{\pi}{2}, \ k \in Z$$

$$\Leftrightarrow \ x \in \bigcup_{k \in Z}\left(k\pi + \frac{7\pi}{12}, k\pi + \frac{3\pi}{4}\right).$$

4. $\dfrac{\sqrt{3}\operatorname{ctg} x + 1}{\sqrt{3} - \operatorname{ctg} x} > \sqrt{3}, \ x \in R \setminus \left(\{k\pi : k \in Z\} \cup \left\{k\pi + \dfrac{\pi}{6} : k \in Z\right\}\right) \ \Leftrightarrow$

$$\operatorname{ctg}\left(\frac{\pi}{6} - x\right) > \sqrt{3} \ \Leftrightarrow \ k\pi < \frac{\pi}{6} - x < k\pi + \frac{\pi}{6} \ k \in Z$$

$$\Leftrightarrow \ x \in \bigcup_{k \in Z}\left(k\pi, k\pi + \frac{\pi}{6}\right).$$

5. $\arcsin x \le \arccos x, \ x \in [-1,1]$

This inequation is equivalent through the agency of the basic identity $\arcsin x + \arccos x = \dfrac{\pi}{2}, \ \forall x \in [-1,1]$ with $\arccos x \ge \dfrac{\pi}{4}$ and, because the function $\arccos$ is strictly decreasing on $[-1,1]$, the solution is $x \in \left[-1, \dfrac{\sqrt{2}}{2}\right]$

6. $\operatorname{arctg} x > \operatorname{arcctg} x, \ x \in R$

By virtue of the basic identity $\operatorname{arctg} x + \operatorname{arcctg} x = \dfrac{\pi}{2}, \ \forall x \in R$, the inequation becomes: $\operatorname{arctg} x > \dfrac{\pi}{4}$ and as the function arctg is strictly increasing on R, we conclude that $x \in (1, +\infty)$

7. $\dfrac{\sin x + \cos x}{\sin x - \cos x} < 0 \iff \dfrac{\sin\left(x + \dfrac{\pi}{4}\right)}{\sin\left(x - \dfrac{\pi}{4}\right)} < 0 \iff$

$$x \in \bigcup_{k \in Z}\left(k\pi - \frac{\pi}{4},\ k\pi\right) \cup \left(k\pi, k\pi + \frac{\pi}{4}\right).$$

8. $\dfrac{\cos x}{1 - \cos 2x} > 0 \iff \dfrac{\cos x}{2\left(1 - \cos^2 x\right)} > 0 \iff$

$$x \in \bigcup_{k \in Z}\left(2k\pi - \frac{\pi}{2},\ 2k\pi\right) \cup \left(2k\pi, 2k\pi + \frac{\pi}{2}\right)$$

9. $8\sin^4 x - 5\sin^2 x + \sin^2 3x \le 0 \iff$

$$\sin^2 x + \sin^2 3x \le 2\sin x\left(3\sin x - 4\sin^3 x\right) \iff \left(\sin x - \sin 3x\right)^2 \le 0 \iff$$

$$\iff \sin x = \sin 3x \iff \sin x = 0 \text{ or } \sin x = \pm\frac{\sqrt{3}}{2} \iff$$

$$\iff x \in \left\{k\pi : k \in Z\right\} \cup \left\{k\pi + (-1)^{k+1}\cdot\frac{\pi}{3} : k \in Z\right\} \cup \left\{k\pi + (-1)^k\frac{\pi}{3} : k \in Z\right\}.$$

10. $\arcsin(2 - x) < \arcsin 2x$. The existence conditions are:

$$\begin{cases} -1 \le 2 - x \le 1 \\ -1 \le 2x \le 1 \end{cases} \iff \begin{cases} x \in [1,3] \\ x \in \left[-\dfrac{1}{2}, \dfrac{1}{2}\right] \end{cases} \iff x \in \left[-\dfrac{1}{2}, \dfrac{1}{2}\right] \cap [1,3] = \phi.$$

5.7. SYSTEMS OF INEQUATIONS

1. $-\sqrt{3} \le \operatorname{tg} 3x \le \sqrt{3} \qquad \iff \qquad -\operatorname{tg}\dfrac{\pi}{3} \le \operatorname{tg} 3x \le \operatorname{tg}\dfrac{\pi}{3} \qquad \iff$

$$3x \in \bigcup_{k \in Z}\left[k\pi - \frac{\pi}{3}, k\pi + \frac{\pi}{3}\right]$$

$$\Leftrightarrow \; x \in \bigcup_{k \in Z} \left[\frac{k\pi}{3} - \frac{\pi}{9}, \frac{k\pi}{3} + \frac{\pi}{9} \right].$$

2. $\begin{cases} 2(1-\cos x) \geq \sin^2 x \\ \sin^4 x + \cos^4 x \geq \dfrac{5}{8} \end{cases} \Leftrightarrow \begin{cases} 4\sin^2 \dfrac{x}{2} \geq 4\sin^2 \dfrac{x}{2}\cos^2 \dfrac{x}{2} \\ \left(\sin^2 x + \cos^2 x\right)^2 - 2\sin^2 x\cos^2 x \geq \dfrac{5}{8} \end{cases} \Leftrightarrow$

$$\Leftrightarrow \begin{cases} 1 \geq \cos^2 \dfrac{x}{2} \\ \dfrac{3}{8} \geq \dfrac{\sin^2 2x}{2} \end{cases} \Leftrightarrow \begin{cases} x \in R \\ \left|\sin 2x\right| \leq \dfrac{\sqrt{3}}{2} \end{cases} \Leftrightarrow$$

$$\Leftrightarrow x \in \bigcup_{k \in Z} \left[2k\pi, 2k\pi + \frac{\pi}{6} \right] \cup \left[2k\pi + \frac{\pi}{3}, 2k\pi + \frac{2\pi}{3} \right] \cup \left[2k\pi + \frac{5\pi}{6}, 2k\pi + \pi \right]$$

3. $\begin{cases} \arcsin(1-x) > \arcsin 2x \\ \arccos(1+x) \leq \arccos(1-x) \end{cases}.$

The conditions of existence for the expressions

lead to $\left|1-x\right|, \left|1+x\right| \left|2x\right| \leq 1$ with the final solution $x = 0$.

4. $\begin{cases} \left|\sin x + \cos x\right| \leq 1 \\ \left|\sin x - \cos x\right| \leq 1 \end{cases} \Leftrightarrow \begin{cases} \left|\cos\left(x - \dfrac{\pi}{4}\right)\right| \leq \dfrac{1}{\sqrt{2}} \\ \left|\cos\left(x + \dfrac{\pi}{4}\right)\right| \leq \dfrac{1}{\sqrt{2}} \end{cases} \Leftrightarrow$

$$\Leftrightarrow k\pi + \frac{\pi}{4} \leq x - \frac{\pi}{4} \leq k\pi + \frac{3\pi}{4}, \; k \in Z$$

$$q\pi + \frac{\pi}{4} \leq x + \frac{\pi}{4} \leq q\pi + \frac{3\pi}{4}, \quad q \in Z \;\Leftrightarrow\; x = h\pi + \frac{\pi}{2}, \; h \in Z.$$

5. $\begin{cases} \left|\operatorname{ctg} x\right| \geq 1 \\ \operatorname{tg} x > \dfrac{1}{\sqrt{3}} \end{cases} \Leftrightarrow \begin{cases} x \in \bigcup_{k \in Z} \left[k\pi - \dfrac{\pi}{4}, k\pi \right) \cup \left(k\pi, k\pi + \dfrac{\pi}{4} \right] \\ x \in \bigcup_{k \in Z} \left(k\pi + \dfrac{\pi}{6}, k\pi + \dfrac{\pi}{2} \right) \end{cases} \Leftrightarrow$

$$\Leftrightarrow x \in \bigcup_{k \in Z}\left(k\pi + \frac{\pi}{4},\ k\pi + \frac{\pi}{6}\right)$$

6. $\begin{cases} \arcsin 4x > -\dfrac{\pi}{6} \\[2mm] \arccos 5x \le \dfrac{\pi}{4} \end{cases} \Leftrightarrow \begin{cases} -\dfrac{1}{2} \le 4x \le 1 \\[2mm] \dfrac{\sqrt{2}}{2} \le 5x \le 1 \end{cases} \Leftrightarrow\ x \in \left[\dfrac{\sqrt{2}}{10}, \dfrac{1}{5}\right].$

7. $\begin{cases} \operatorname{arctg} 3x \le \dfrac{\pi}{4} \\[2mm] \operatorname{arcctg}(2x+1) > \dfrac{\pi}{2} \end{cases} \Leftrightarrow \begin{cases} 3x \le 1 \\[1mm] 2x+1 < 0 \end{cases} \Leftrightarrow\ x \in \left(-\infty, -\dfrac{1}{2}\right).$

8. $\begin{cases} \arcsin x \ge \arccos x \\[1mm] \operatorname{arctg} x < \operatorname{arcctg} x \end{cases} \Leftrightarrow \begin{cases} \arccos x \le \dfrac{\pi}{4} \\[2mm] \operatorname{arctg} x < \dfrac{\pi}{4} \end{cases} \Leftrightarrow \begin{cases} x \in \left[\dfrac{\sqrt{2}}{2}, 1\right] \\[2mm] x \in (-\infty, 1) \end{cases} \Leftrightarrow$

$$x \in \left[\frac{\sqrt{2}}{2}, 1\right).$$

9. $\begin{cases} \arccos x + \arcsin(1-x) > \dfrac{\pi}{2} \\[2mm] |\operatorname{tg} x| \le 1 \end{cases} \Leftrightarrow \begin{cases} |x|,\ |1-x| \le 1 \\[1mm] \arccos x > \arccos(1-x) \\[1mm] x \in \bigcup_{k \in Z}\left[k\pi - \dfrac{\pi}{4}, k\pi + \dfrac{\pi}{4}\right] \end{cases} \Leftrightarrow$

$$\Leftrightarrow \begin{cases} x \in [0,1] \\[1mm] x < \dfrac{1}{2} \\[2mm] x \in \bigcup_{k \in Z}\left[k\pi - \dfrac{\pi}{4}, k\pi + \dfrac{\pi}{4}\right] \end{cases} \Leftrightarrow\ x \in \left[0, \frac{1}{2}\right).$$

10.
$$
\begin{cases}
\operatorname{arctg} x + 2\operatorname{arctg}\dfrac{1}{x} > \dfrac{\pi}{2} \\[2ex]
\operatorname{arctg}\dfrac{1}{x} + \operatorname{arctg} x > \dfrac{\pi}{4} \\[2ex]
x \geq \dfrac{1}{5}
\end{cases}
$$

Due to $\operatorname{arctg} x + \operatorname{arctg}\dfrac{1}{x} = \begin{cases} \dfrac{\pi}{2}, x > 0 \\[2ex] -\dfrac{\pi}{2}, x < 0 \end{cases}$, the system becomes:

$$
\begin{cases}
\operatorname{arctg}\dfrac{1}{x} > 0 \\[2ex]
\operatorname{arctg} x > -\dfrac{\pi}{4} \\[2ex]
x \geq \dfrac{1}{5}
\end{cases}
\Leftrightarrow
\begin{cases}
x \in (0,+\infty) \\[2ex]
x \in (-1,+\infty) \\[2ex]
x \in \left[\dfrac{1}{5},+\infty\right)
\end{cases}
\Leftrightarrow
x \in \left[\dfrac{1}{5},+\infty\right).
$$

5.8. THE TRIGONOMETRIC EXPRESSIONS OF COMPLEX NUMBERS

As we have mentioned at the end of section **A** of the paragraph 1.3. in the first chapter, the set $C = R^2 = \{(a,b): a,b \in R\}$ was acknowledged as *set of complex numbers* in 1837 with the help of the combined contributions of the mathematicians Rowan William Hamilton, Louis Augustin Cauchy, Richard Wilhelm Dedekind and Karl Wilhelm Weierstrass.

The following operations $+, \cdot : C \times C \to C$ defined by $(a_1,b_1)+(a_2,b_2)=(a_1+a_2,b_1+b_2)$, $(a_1,b_1)\cdot(a_2,b_2)=(a_1 a_2 - b_1 b_2, a_1 b_2 + a_2 b_1)$, $\forall z_1 = (a_1,b_1)$, $z_2 = (a_2,b_2) \in C$ induce on C o an algebraic structure of commutative field under the convention that any couple $(a,0)(a \in R)$ is identified with the real number a and one denotes $i = (0,1)$, then the precedent operations support the customary description of complex numbers $z = (a,b)$ by $z = a + ib$ or the algebraic equivalent,

$$z = \sqrt{a^2 + b^2}\left(\frac{a}{\sqrt{a^2+b^2}} + i\frac{b}{\sqrt{a^2+b^2}}\right).$$

Because $\left(\dfrac{a}{\sqrt{a^2+b^2}}\right)^2 + \left(\dfrac{b}{\sqrt{a^2+b^2}}\right)^2 = 1$, there is $\varphi \in [0, 2\pi]$ called the *argument* of the number z, so that $\cos\varphi = \dfrac{a}{\sqrt{a^2+b^2}}$ and $\sin\varphi = \dfrac{b}{\sqrt{a^2+b^2}}$.

Marking with $|z| = \sqrt{a^2 + b^2}$ the *module* of z, one concludes that

$$z = |z|(\cos\varphi + i\sin\varphi), \quad z_1 = a_1 + ib_1 = z_2 = a_2 + ib_2, \quad (a_1, a_2, b_1, b_2 \in R) \iff$$

$a_1 = a_2$ and $b_1 = b_2$, and *the conjugated* of any complex number $z = a + ib, \ (a, b \in R)$ is $\overline{z} = a - ib$.

The previous algebraic and the trigonometric representation allows also a significative geometric interpretation officially suggested in 1831 by Karl Friedrich Gauss and adjusted for R^2, considered as the two-dimensional real linear space of the vectors, naturally projected in any plane (see Figure 35).

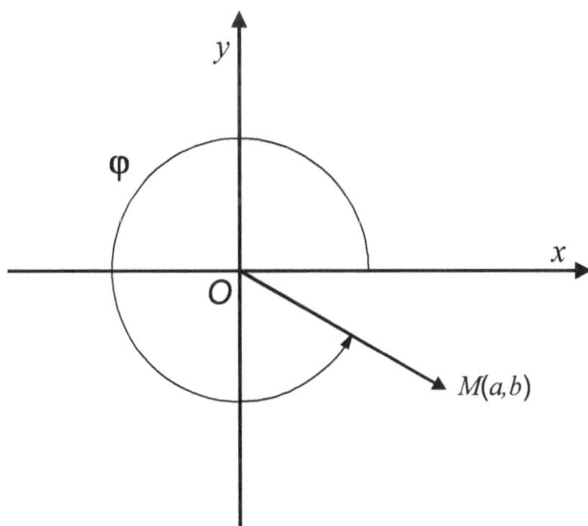

Figure 35: The primary plane projection of the complex numbers

$$\overrightarrow{OM} = a\vec{i} + b\vec{j} = \left\|\overrightarrow{OM}\right\|\left[\left(\cos\varphi\right)\vec{i} + \left(\sin\varphi\right)\vec{j}\right],$$ z being called *the afix*

of the geometric representation $M(a,b)$.

1. $\forall z_1 = |z_1|\left(\cos\varphi_1 + i\sin\varphi_1\right),\ z_2 = |z_2|\left(\cos\varphi_2 + i\sin\varphi_2\right) \Rightarrow$

$$\Rightarrow z_1 z_2 = |z_1||z_2|\left[\cos\left(\varphi_1 + \varphi_2\right) + i\sin\left(\varphi_1 + \varphi_2\right)\right].$$

Thus $\forall z = |z|\left(\cos\varphi + i\sin\varphi\right) \Rightarrow z^r = |z|^r\left(\cos r\varphi + i\sin r\varphi\right),\ \forall r \in Q$
(relation established by Abraham de Moivre (1667-1754) for $r \in Z$ and generalized for the rational exponents by Leonhard Euler (1707-1783)).

2. $\forall z = |z|\left(\cos\varphi + i\sin\varphi\right),\ n \in N^*,\ n \geq 2 \Rightarrow$

$$\sqrt[n]{z} = \sqrt[n]{|z|}\left(\cos\frac{2k\pi + \varphi}{n} + i\sin\frac{2k\pi + \varphi}{n}\right),\ k = \overline{0, n-1}.$$

3. $|z| = 1 \Leftrightarrow \exists \lambda \in R : z = \dfrac{1 + \lambda \cdot i}{1 - \lambda \cdot i}.$

Indeed, if $|z| = 1$, then $z = \cos\varphi + i\sin\varphi$ and the equality

$\cos\varphi + i\sin\varphi = \dfrac{1 + \lambda \cdot i}{1 - \lambda \cdot i}$ implies that $\lambda = \dfrac{1 - \cos\varphi}{\sin\varphi} = \dfrac{\sin\varphi}{1 + \cos\varphi}$. If

$z = \dfrac{1 + \lambda \cdot i}{1 - \lambda \cdot i}$ with $\lambda \in R$, then $z = \dfrac{1 - \lambda^2}{1 + \lambda^2} + \dfrac{2\lambda}{1 + \lambda^2}i$ and

$$\left(\frac{1 - \lambda^2}{1 + \lambda^2}\right)^2 + \left(\frac{2\lambda}{1 + \lambda^2}\right)^2 = 1.$$

4. The condition that characterizes the equality of the complex numbers together with the first two applications prove that the equation in C

$$\left(\frac{1 + iz}{1 - iz}\right)^n = \cos\varphi + i\sin\varphi,\ n \in N^*,\ n \geq 2,\ \varphi \in [0, 2\pi]$$ can be

written under the equivalent form

$$\frac{1+iz}{1-iz} = \cos\frac{2k\pi+\varphi}{n} + i\frac{2k\pi+\varphi}{n}, \; k = \overline{0, n-1}, \text{ that is, it has the}$$

following solutions $z_k = \text{tg}\dfrac{k\pi+\dfrac{\varphi}{2}}{n}, \; k = \overline{0, n-1}$.

5.   If $z + \dfrac{1}{z} = 2\cos\varphi,$ then $z = \cos\varphi \pm i\sin\varphi,$

$$z^r = \cos r\varphi \pm i\sin r\varphi,$$

$$z^{-r} = \cos r\varphi \mp i\sin r\varphi, \text{ so } z^r + \frac{1}{z^r} = 2\cos r\varphi, \; \forall r \in Q \text{ and}$$

$$\sum_{k=1}^{n}\left(z^k + \frac{1}{z^k}\right) = \frac{2\cos\dfrac{n+1}{2}\varphi\sin\dfrac{n\varphi}{2}}{\sin\dfrac{\varphi}{2}}.$$

6. The roots of the equation $z^n = 1, \; n \in N^*, \; n \geq 2$ are

$$z_k = \cos\frac{2k\pi}{n} + i\sin\frac{2k\pi}{n}, \; k = \overline{0, n-1} \text{ with } \sum_{k=0}^{n-1} z_k = 0.$$

Therefore, $\displaystyle\sum_{k=0}^{n-1}\cos\frac{2k\pi}{n} = \sum_{k=0}^{n-1}\sin\frac{2k\pi}{n} = 0.$

7. The equation $z^{2n} = 1, \; n \in N^*$ has the roots $1, -1,$

$$z_k = \cos\frac{k\pi}{n} + i\sin\frac{k\pi}{n} \text{ and } \overline{z_k} = \cos\frac{k\pi}{n} - i\sin\frac{k\pi}{n}, \; k = \overline{1, n-1},$$

so

$$\frac{z^{2n}-1}{z^2-1} = 1 + z^2 + z^4 + \dots + z^{2n-2} = \prod_{k=1}^{n-1}(z-z_k)(z-\overline{z_k}), \; \forall z \in C \setminus \{-1, 1\}.$$

Thus, $1 + z^2 + z^4 + \dots + z^{2n-2} = \displaystyle\prod_{k=1}^{n-1}\left(z^2 - 2z\cos\frac{k\pi}{n} + 1\right), \; \forall z \in C$

from   which   for $z = \pm1$ one   obtains

$$\prod_{k=1}^{n-1}\sin\frac{k\pi}{2n}=\prod_{k=1}^{n-1}\cos\frac{k\pi}{2n}=\frac{\sqrt{n}}{2^{n-1}}$$ which implies that

$$\prod_{k=1}^{n-1}\sin\frac{k\pi}{n}=\frac{n}{2^{n-1}}$$ and, considering $z=i$ it results that

$$\prod_{k=1}^{n-1}\cos\frac{k\pi}{n}=\begin{cases}0,\ n=2k\\[2mm]\dfrac{(-1)^{k}}{2^{2k}},\ n=2k+1,\ k\in N^{*}\end{cases}$$

8. The formula of Moivre allows the writing of the next equation in R:

$$\prod_{k=1}^{n}\left(\cos kx+i\sin kx\right)=1$$ under the form

$$\cos\frac{n(n+1)}{2}x+i\sin\frac{n(n+1)}{2}x=1 \iff \begin{cases}\cos\dfrac{n(n+1)}{2}x=1\\[3mm]\sin\dfrac{n(n+1)}{2}x=0\end{cases} \iff$$

$$x=\frac{4k\pi}{n(n+1)},\ k\in Z\ .$$

9. If one considers $\displaystyle S_1=\sum_{k=1}^{n}\sin kx,$

$\displaystyle S_2=\sum_{k=1}^{n}\cos kx,\ x\in R,\ n\in N^{*}$, then the formula of Moivre

combined with the identity $\displaystyle\sum_{k=1}^{n}z^{k}=\begin{cases}z\cdot\dfrac{z^{n}-1}{z-1},\ z\neq1\\[3mm]n,\ z=1\end{cases}$ implies

that $S_2+iS_1=\dfrac{\sin\dfrac{nx}{2}}{\sin\dfrac{x}{2}}\left(\cos\dfrac{n+1}{2}x+i\sin\dfrac{n+1}{2}x\right)$

Consequently,
$$S_1 = \frac{\sin\dfrac{nx}{2}\cdot\sin\dfrac{n+1}{2}x}{\sin\dfrac{x}{2}},$$

$$S_2 = \frac{\sin\dfrac{nx}{2}\cdot\cos\dfrac{n+1}{2}x}{\sin\dfrac{x}{2}}.$$

10. $\quad z^{2n+1}=1 \iff (z-1)\left(z^{2n}+z^{2n-1}+...+z+1\right)=0 \iff$

$$z_k = \cos\frac{2k\pi}{2n+1}+i\sin\frac{2k\pi}{2n+1},\ k=\overline{0,2n}.$$

Taking $x=\dfrac{1}{1+z}$ in the equation $z^{2n}+z^{2n-1}+...+z+1=0$,

from the sum of the roots of the equation obtained in

this manner we deduce that $\displaystyle\sum_{k=1}^{2n}\frac{1}{1+z_k}=n$ and the

change of variables $y=\dfrac{1}{z}$ made in the same equation

leads to $\displaystyle\sum_{k=1}^{2n}\frac{1}{z_k}=-1.$

Consequently,
$$\sum_{k=1}^{2n}\frac{1-\overline{z}_k}{1+z_k}=2\sum_{k=1}^{2n}\frac{1}{1+z_k}-\sum_{k=1}^{2n}\frac{1}{z_k}=2n+1$$

because $z_k\cdot\overline{z}_k=1,\ \forall k=\overline{0,2n}.$

5.9. SOME GEOMETRICAL APPLICATIONS

For any $\triangle ABC$ we write $\left\|\overrightarrow{AB}\right\|=c,\ \left\|\overrightarrow{BC}\right\|=a,\ \left\|\overrightarrow{AC}\right\|=b,$

$\widehat{CAB}=\hat{A},\quad \widehat{ABC}=\hat{B},\ \widehat{BCA}=\hat{B}.$ (see Figure 36).

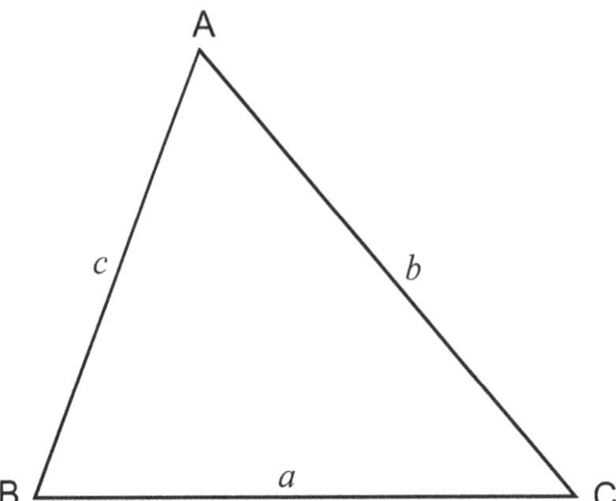

Figure 36: A triangle projection

Because $\overrightarrow{BC} = \overrightarrow{AC} - \overrightarrow{AB}$, using the properties of the scalar products, we obtain:

$$\left\|\overrightarrow{BC}\right\|^2 = \left\langle\overrightarrow{BC},\overrightarrow{BC}\right\rangle = \left\|\overrightarrow{AC}\right\|^2 + \left\|\overrightarrow{AB}\right\|^2 - 2\left\|\overrightarrow{AB}\right\|\cdot\left\|\overrightarrow{AC}\right\|\cos\widehat{CAB}.$$

Thus,

$$\begin{cases} a^2 = b^2 + c^2 - 2bc\cos\hat{A} \\ b^2 = a^2 + c^2 - 2ac\cos\hat{B} \\ c^2 = a^2 + b^2 - 2ab\cos\hat{C} \end{cases}$$

(the theorem of the Cosinus or, equivalently, the generalized theorem of Pythagoras).

By considering that the circumscribed circle to any

non-degenerated triangle $\triangle ABC$ is centered in O having the radius R,

and D being diametrically opposite to C, then, $\widehat{CDB} \equiv \widehat{CAB} = \hat{A}$

$\left\|DC\right\| = 2R,\ m(\widehat{DBC}) = \dfrac{\pi}{2},\ \sin\hat{A} = \sin\widehat{CDB} = \dfrac{a}{2R}$ (see Figure 37).

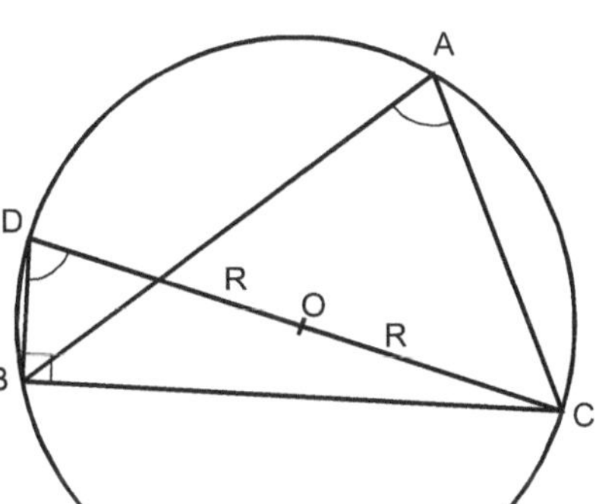

Figure 37: The sine function and the circumscribed circle

Similar arguments justifies the relations $\sin \hat{B} = \dfrac{b}{2R}$ and $\sin \hat{C} = \dfrac{c}{2R}$.

Concluding,

$$\frac{a}{\sin \hat{A}} = \frac{b}{\sin \hat{B}} = \frac{c}{\sin \hat{C}} = 2R, \ \forall \Delta ABC \text{ non-degenerated } \textbf{\textit{(the theorem}}$$

of the sinus).

Based on this theorem combined with the considerations of Chapter IV regarding the frequent trigonometric identities, convenient to each of the next models, it results that

$$\frac{a-b}{a+b} = \frac{2R\left(\sin \hat{A} - \sin \hat{B}\right)}{2R\left(\sin \hat{A} + \sin \hat{B}\right)} = \frac{tg\dfrac{\hat{A}-\hat{B}}{2}}{tg\dfrac{\hat{A}+\hat{B}}{2}}, \ \frac{b-c}{b+c} = \frac{tg\dfrac{\hat{B}-\hat{C}}{2}}{tg\dfrac{\hat{B}+\hat{C}}{2}},$$

$$\frac{c-a}{c+a} = \frac{tg\dfrac{\hat{C}-\hat{A}}{2}}{tg\dfrac{\hat{C}+\hat{A}}{2}},$$

(the theorem of the tangents)

$$a = b\cos \hat{C} + c\cos \hat{B}, \ b = c\cos \hat{A} + a\cos \hat{C}, \ c = a\cos \hat{B} + b\cos \hat{A},$$

(the theorem of the projections)

respectively,

$$\frac{a\pm b}{c}=\frac{\cos(\sin)\dfrac{\hat{A}-\hat{B}}{2}}{\sin(\cos)\dfrac{\hat{C}}{2}},\quad \frac{b\pm c}{a}=\frac{\cos(\sin)\dfrac{\hat{B}-\hat{C}}{2}}{\sin(\cos)\dfrac{\hat{A}}{2}},\quad \frac{c\pm a}{b}=\frac{\cos(\sin)\dfrac{\hat{C}-\hat{A}}{2}}{\sin(\cos)\dfrac{\hat{B}}{2}}$$

(the formulae of Karl Mollweide) in which "+" corresponds to the first application of the first trigonometric function by ignoring the succesor, and "-" corresponds to the neglecting of the first function, correspondingly managing the simultaneous circular permutations $a\to b\to c\to a$ and $\hat{A}\to\hat{B}\to\hat{C}\to\hat{A}$.

The same arguments associated with Heron's formula which express the area S for any non-degenerated $\triangle ABC$ by

$$S=\sqrt{p(p-a)(p-b)(p-c)},\ \text{ where }\ p=\frac{a+b+c}{2}\ \text{ justifies the}$$

next relations:

$$\sin\frac{\hat{A}}{2}=\sqrt{\frac{(p-b)(p-c)}{bc}},\qquad\qquad \sin\frac{\hat{B}}{2}=\sqrt{\frac{(p-c)(p-a)}{ac}},$$

$$\sin\frac{\hat{C}}{2}=\sqrt{\frac{(p-a)(p-b)}{ab}},$$

$$\cos\frac{\hat{A}}{2}=\sqrt{\frac{p(p-a)}{bc}},\ \ \cos\frac{\hat{B}}{2}=\sqrt{\frac{p(p-b)}{ac}},\ \ \cos\frac{\hat{C}}{2}=\sqrt{\frac{p(p-c)}{ab}},\ \text{thus}$$

$$tg\frac{\hat{A}}{2}=\sqrt{\frac{(p-b)(p-c)}{p(p-a)}},\qquad\qquad tg\frac{\hat{B}}{2}=\sqrt{\frac{(p-c)(p-a)}{p(p-b)}}\qquad \text{and}$$

$$tg\frac{\hat{C}}{2}=\sqrt{\frac{(p-a)(p-b)}{p(p-c)}}.$$

But, $S=\dfrac{ab\sin\hat{C}}{2}=\dfrac{bc\sin\hat{A}}{2}=\dfrac{ac\sin\hat{B}}{2}=\dfrac{abc}{4R}.$

At the same time, if h_a, h_b, h_c designates the lengths of the heights corresponding to the sides $|BC|,|AC|,|AB|,\ S=\dfrac{a\cdot h_a}{2}=\dfrac{b\cdot h_b}{2}=\dfrac{c\cdot h_c}{2}.$

Therefore, $a = \dfrac{2S}{h_a}$, $b = \dfrac{2S}{h_b}$, $c = \dfrac{2S}{h_c}$, $p = S\left(\dfrac{1}{h_a} + \dfrac{1}{h_b} + \dfrac{1}{h_c}\right)$,

$$p - a = S\left(\dfrac{1}{h_b} + \dfrac{1}{h_c} - \dfrac{1}{h_a}\right), \qquad\qquad p - b = S\left(\dfrac{1}{h_a} + \dfrac{1}{h_c} - \dfrac{1}{h_b}\right),$$

$$p - c = S\left(\dfrac{1}{h_a} + \dfrac{1}{h_b} - \dfrac{1}{h_c}\right) \text{ and Heron's formula is equivalent to}$$

$$S = S^2 \sqrt{\left(\dfrac{1}{h_a} + \dfrac{1}{h_b} + \dfrac{1}{h_c}\right)\left(\dfrac{1}{h_b} + \dfrac{1}{h_c} - \dfrac{1}{h_a}\right)\left(\dfrac{1}{h_a} + \dfrac{1}{h_c} - \dfrac{1}{h_b}\right)\left(\dfrac{1}{h_a} + \dfrac{1}{h_b} - \dfrac{1}{h_c}\right)},$$

that is,

$$S = \dfrac{1}{\sqrt{\left(\dfrac{1}{h_a} + \dfrac{1}{h_b} + \dfrac{1}{h_c}\right)\left(\dfrac{1}{h_b} + \dfrac{1}{h_c} - \dfrac{1}{h_a}\right)\left(\dfrac{1}{h_a} + \dfrac{1}{h_c} - \dfrac{1}{h_b}\right)\left(\dfrac{1}{h_a} + \dfrac{1}{h_b} - \dfrac{1}{h_c}\right)}}.$$

By attaching the inscribed circle of center I and radius r the following description is realized (see Figure 38).

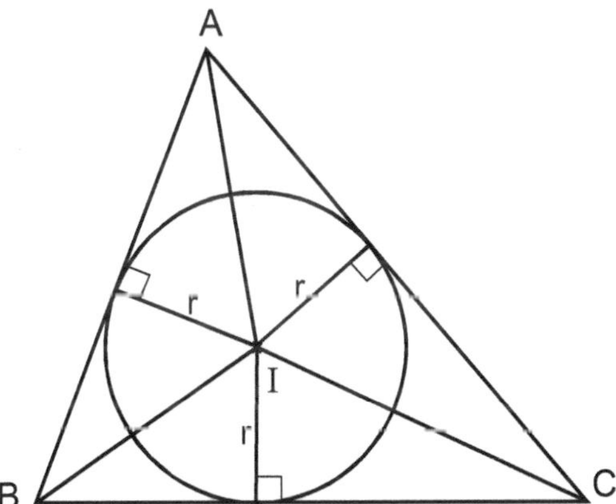

Figure 38: The tangent function and the inscribed circle

which, supported by the previous relations between the tangent of each semi-angle, the lengths of the triangle sides and the identity $r \cdot p = S$ motivates the relations:

$$r = \sqrt{\frac{(p-a)(p-b)(p-c)}{p}} = (p-a)tg\frac{\hat{A}}{2} = (p-b)tg\frac{\hat{B}}{2} = (p-c)tg\frac{\hat{C}}{2} =$$

$$= p\cdot tg\frac{\hat{A}}{2}\cdot tg\frac{\hat{B}}{2}\cdot tg\frac{\hat{C}}{2} = 4R\sin\frac{\hat{A}}{2}\sin\frac{\hat{B}}{2}\sin\frac{\hat{C}}{2}$$

$$p = 4R\cos\frac{\hat{A}}{2}\cos\frac{\hat{B}}{2}\cos\frac{\hat{C}}{2}, \quad p-a = 4R\cos\frac{\hat{A}}{2}\sin\frac{\hat{B}}{2}\sin\frac{\hat{C}}{2},$$

$$p-b = 4R\cos\frac{\hat{B}}{2}\sin\frac{\hat{A}}{2}\sin\frac{\hat{C}}{2}, \quad p-c = 4R\cos\frac{\hat{C}}{2}\sin\frac{\hat{A}}{2}\sin\frac{\hat{B}}{2}.$$

If r_a, r_b, r_c are the radiuses of the exainscribed circles tangent to the sides $|BC|$, $|AC|$ and $|AB|$, then $S = (p-a)r_a = (p-b)r_b = (p-c)r_c$, which together with the adequate previous relations explain the following colligations:

$$r_a = r + atg\frac{\hat{A}}{2}, \quad r_b = r + btg\frac{\hat{B}}{2}, \quad r_c = r + c\cdot tg\frac{\hat{C}}{2}.$$

An important consequence of the theorem of the cosines is:

1. If $A-B-C$ and $O \notin AC$, then

$$\|OA\|^2 \cdot \|BC\| - \|OB\|^2 \cdot \|AC\| + \|OC\|^2 \cdot \|AB\| = \|AB\| \cdot \|BC\| \cdot \|AC\|$$

(Stewart's relation) (see Figure 39).

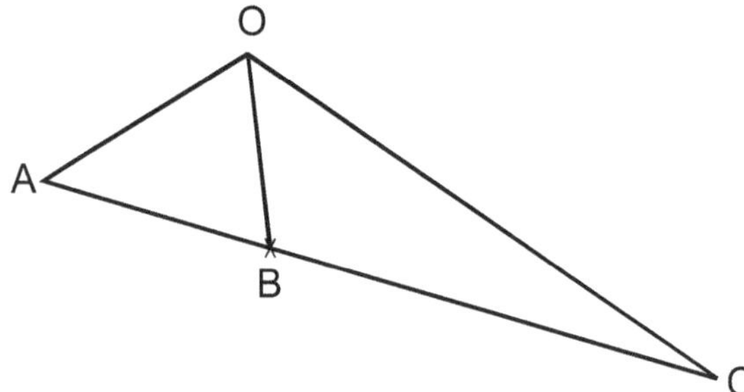

Figure 39: An application of the cosine function

because, according to the theorem of the cosines applied in $\triangle OBA$ and $\triangle OBC$, it follows that

$$\|OA\|^2 = \|OB\|^2 + \|AB\|^2 - 2\|OB\| \cdot \|AB\| \cos\left(\widehat{ABO}\right),$$

$$\|OC\|^2 = \|OB\|^2 + \|BC\|^2 + 2\|OB\| \cdot \|BC\| \cos\left(\widehat{ABO}\right)$$

equality which, by successive multiplication with $\|BC\|$, $\|AB\|$ and the algebraic addition justifies the indicated account. In the particular case when m_a represents the length, the median corresponding to the $|BC|$ side of any non-trivial triangle $\triangle ABC$ and m_b, m_c are the analogous for m_a (see Figure 40).

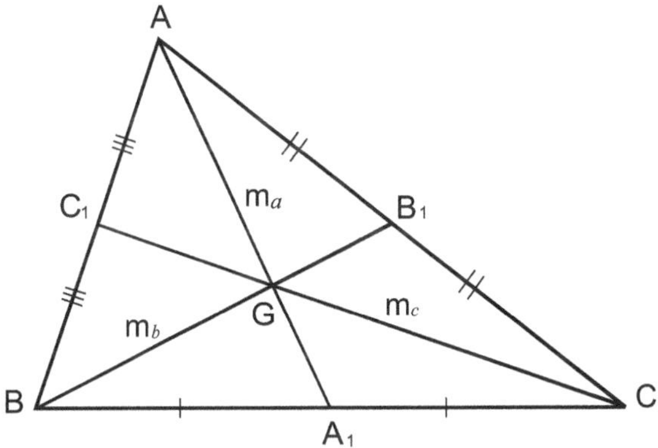

Figure 40: The medians in usual triangles

$$A_1 \in |BC|, \quad |A_1B| \equiv |A_1C|,$$

$$B_1 \in |AC|, \quad |B_1A| \equiv |B_1C|,$$

$$C_1 \in |AB|, \quad |C_1B| \equiv |C_1A|,$$

$$\{G\} = AA_1 \cap BB_1 \cap CC_1,$$

$$\frac{\|GA\|}{\|GA_1\|} = \frac{\|GB\|}{\|GB_1\|} = \frac{\|GC\|}{\|GC_1\|} = 2,$$ the point G r representing *the weight center* of the plane (ABC) plate every time this is *physically homogeneous* and $$m_a^2 = \frac{2\left(b^2 + c^2\right) - a^2}{4},$$ $$m_b^2 = \frac{2\left(a^2 + c^2\right) - b^2}{4},$$

$$m_c^2 = \frac{2\left(a^2 + b^2\right) - c^2}{4},$$ *(the theorem of the median).*

2. The geometric place of all the points in any Euclidean plane (space) with the constant sum of the squares of the distances between two fix distinct, proper points is a circle (a sphere).

Indeed, if A, B are the fix points, $T \in |AB|$, $|TA| \equiv |TB|$ and $\|MA\|^2 + \|MB\|^2 = k$, then, according to the theorem of the median,

$$\|MT\|^2 = \frac{2\left(\|MA\|^2 + \|MB\|^2\right) - \|AB\|^2}{4} = \frac{2k - \|AB\|^2}{4} \quad \text{(see Figure 41).}$$

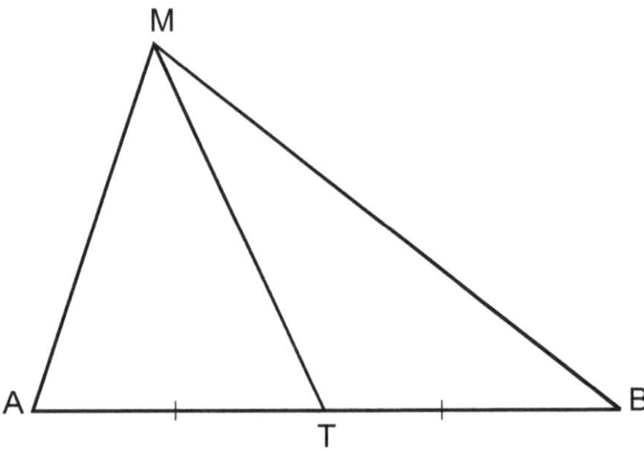

Figure 41: Another application of the medians

For possibility, the condition $k \geq \dfrac{\|AB\|^2}{2}$ is imposed, and M belongs to the circle (sphere) of center T and radius $\dfrac{\sqrt{2k - \|AB\|^2}}{2}$.

3. The convex quadrilateral of maximum area, realised by 4 uncommon segments is that inscriptible (when it exists) (see Figure 42).

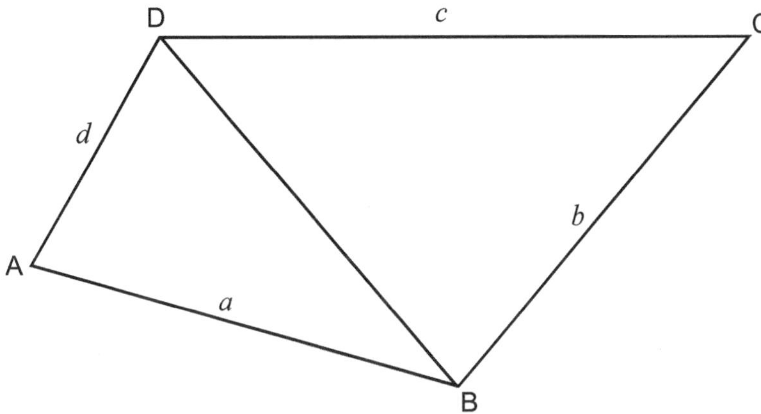

Figure 42: Maximum of convex quadrilaterals

In order to justify, we write, $\|AB\| = a,\ \|BC\| = b,\ \|CD\| = c,\ \|AD\| = d$,
$p = \dfrac{a+b+c+d}{2}$ and by S the area of the quadrilateral ABCD $ABCD$.

Then, $S = \dfrac{ad \sin \widehat{DAB}}{2} + \dfrac{bc \sin \widehat{BCD}}{2}$, and the theorem of the cosines applied for $|BD|$ in $\triangle ABD$ and $\triangle DBC$ indicates that

$$\|BD\|^2 = a^2 + d^2 - 2ad \cos \widehat{DAB} = b^2 + c^2 - 2bc \cos \widehat{BCD}.$$

Thus,

$$ad \sin \widehat{DAB} + bc \sin \widehat{BCD} = 2S$$

and

$$2ad \cos \widehat{DAB} - 2bc \cos \widehat{BCD} = a^2 + d^2 - b^2 - c^2.$$

The simultaneous square raising and addition in this position implies that

$$S = \sqrt{(p-a)(p-b)(p-c)(p-d) - abcd \cos^2 \frac{\widehat{DAB} + \widehat{BCD}}{2}} \quad \text{with maxim}$$

value for $m\left(\widehat{DAB}\right) + m\left(\widehat{BCD}\right) = \pi$.

In this situation, $S = \sqrt{(p-a)(p-b)(p-c)(p-d)}$ expression that generalizes Heron's formula, being mentioned by Brahmagupta in the 7th century and anticipated by Archimedes. The lengths of the diagonals d_1 and d_2 of the inscriptible quadrilateral is expressed in relation to the sides lengths by:

$$d_1 = \sqrt{\frac{(ab+cd)(ac+bd)}{ad+bc}}, \quad d_2 = \sqrt{\frac{(ad+bc)(ac+bd)}{ab+cd}} \ .$$

4. In any non-degenerated triangle $\Delta ABC,\ m\left(\widehat{A}\right)+m\left(\widehat{B}\right)+m\left(\widehat{C}\right)=\pi$

$$\Rightarrow\ \cos\left(\widehat{A}+\widehat{B}+\widehat{C}\right)=-1$$

$$\Leftrightarrow$$

$$\cos\widehat{A}\sin\widehat{B}\sin\widehat{C}+\cos\widehat{B}\sin\widehat{C}\sin\widehat{A}+\cos\widehat{C}\sin\widehat{A}\sin\widehat{B}=1+\cos\widehat{A}\cos\widehat{B}\cos\widehat{C}$$

which justifies the coincidence

$$S\left(ctg\widehat{A}+ctg\widehat{B}+ctg\widehat{C}\right)=2R^2\left(1+\cos\widehat{A}\cos\widehat{B}\cos\widehat{C}\right).$$

5. $m\left(\widehat{A}\right)+m\left(\widehat{B}\right)+m\left(\widehat{C}\right)=\pi\ \Rightarrow\ \not\exists tg\dfrac{\widehat{A}+\widehat{B}+\widehat{C}}{2}\ \Leftrightarrow$

$$\Leftrightarrow\ tg\frac{\widehat{A}}{2}\cdot tg\frac{\widehat{B}}{2}+tg\frac{\widehat{B}}{2}\cdot tg\frac{\widehat{C}}{2}+tg\frac{\widehat{C}}{2}\cdot tg\frac{\widehat{A}}{2}=1\ \Rightarrow$$

$$\Rightarrow\frac{1}{27}=\left(\frac{tg\dfrac{\widehat{A}}{2}\cdot tg\dfrac{\widehat{B}}{2}+tg\dfrac{\widehat{B}}{2}\cdot tg\dfrac{\widehat{C}}{2}+tg\dfrac{\widehat{C}}{2}\cdot tg\dfrac{\widehat{A}}{2}}{3}\right)^3\geq\left(tg\frac{\widehat{A}}{2}\cdot tg\frac{\widehat{B}}{2}\cdot tg\frac{\widehat{C}}{2}\right)^2\ \Leftrightarrow$$

$$\Leftrightarrow\ tg\frac{\widehat{A}}{2}\cdot tg\frac{\widehat{B}}{2}\cdot tg\frac{\widehat{C}}{2}\leq\frac{1}{3\sqrt{3}},\quad \text{with equality if and only if the}$$

triangle ΔABC is equilateral.

6. Because the sine function is strictly concave on the interval $[0,\pi]$, in any uncommon triangle

$$ABC,\ \sin\widehat{A}+\sin\widehat{B}+\sin\widehat{C}\leq 3\sin\frac{\widehat{A}+\widehat{B}+\widehat{C}}{3}=\frac{3\sqrt{3}}{2}\ ;\ \text{thus, applying}$$

the usual inequality between the arithmetical and the

geometrical means it follows that $\sin\widehat{A}\cdot\sin\widehat{B}\cdot\sin\widehat{C}\le\dfrac{3\sqrt{3}}{8}$, and the strictly concavity of the cosine function in the interval $\left[0,\dfrac{\pi}{2}\right]$ implies that

$$\left|\cos\widehat{A}+\cos\widehat{B}+\cos\widehat{C}\right|\le\frac{\left|\cos\widehat{A}+\cos\widehat{B}\right|}{2}+\frac{\left|\cos\widehat{B}+\cos\widehat{C}\right|}{2}+\frac{\left|\cos\widehat{C}+\cos\widehat{A}\right|}{2}\le$$

$$\le\left|\cos\frac{\widehat{A}+\widehat{B}}{2}\right|+\left|\cos\frac{\widehat{B}+\widehat{C}}{2}\right|+\left|\cos\frac{\widehat{C}+\widehat{A}}{2}\right|\le 3\cos\frac{\widehat{A}+\widehat{B}+\widehat{C}}{3}=\frac{3}{2},$$

in both inequalities the equality occurs only when $\triangle ABC$ is equilateral.

7. With the indicated notations, $R(a+c)\ge b\sqrt{ac}$ because using the theorem of the sines the inequality becomes $\sin\widehat{A}+\sin\widehat{C}\ge 2\sin\widehat{B}\sqrt{\sin\widehat{A}\cdot\sin\widehat{C}}$ valid, since,

$$\sin\widehat{A}+\sin\widehat{C}\ge 2\sqrt{\sin\widehat{A}\sin\widehat{C}}\ge 2\sin\widehat{B}\sqrt{\sin\widehat{A}\sin\widehat{C}}\ \text{ with equality for}$$

$$m\left(\widehat{B}\right)=\frac{\pi}{2},\ \ m\left(\widehat{A}\right)=m\left(\widehat{C}\right)=\frac{\pi}{4}.$$

8. The strictly concavity of the sine function on the interval $[0,\pi]$ combined with the inequality of means mentioned above show that in any uncommon triangle $\triangle ABC$,

$$\sin\frac{\widehat{A}}{2}\cdot\sin\frac{\widehat{B}}{2}\cdot\sin\frac{\widehat{C}}{2}\le\left(\frac{\sin\dfrac{\widehat{A}}{2}+\sin\dfrac{\widehat{B}}{2}+\sin\dfrac{\widehat{C}}{2}}{3}\right)^{3}\le\left(\sin\frac{\pi}{6}\right)^{3}=\frac{1}{8},$$

together with the same inequality leads to:

$$\cos A\cdot\cos B\cdot\cos C\le\left(\frac{\cos A+\cos B+\cos C}{3}\right)^{3}\le\frac{1}{8}\ \text{ the equal sign}$$

occuring only for the equilateral triangles.

9. Because $\sin 2\hat{A} + \sin 2\hat{B} + \sin 2\hat{C} = 4\sin\hat{A}\sin\hat{B}\sin\hat{C}$ according to the basic identities and $tg\hat{A} + tg\hat{B} + tg\hat{C} = tg\hat{A}\cdot tg\hat{B}\cdot tg\hat{C},\ \ \forall\Delta ABC$ non-degenerated , according to the theorem of the sines reunited with the relation $S = 2R^2\sin\hat{A}\sin\hat{B}\sin\hat{C}$ we conclude that

$$a^2 tg\hat{A} + b^2 tg\hat{B} + c^2 tg\hat{C} = 4R^2\left(tg\hat{A}\cdot tg\hat{B}\cdot tg\hat{C} - 2\sin\hat{A}\sin\hat{B}\sin\hat{C}\right) =$$

$$= 2S\left(\sec\hat{A}\sec\hat{B}\sec\hat{C} - 2\right)$$

10. $m\left(\hat{A}\right) + m\left(\hat{B}\right) + m\left(\hat{C}\right) = \pi \ \Rightarrow\ \not\exists\ ctg\left(\hat{A} + \hat{B} + \hat{C}\right)\ \Leftrightarrow$

$\Leftrightarrow\quad ctg\hat{A}\cdot ctg\hat{B} + ctg\hat{B}\cdot ctg\hat{C} + ctg\hat{C}\cdot ctg\hat{A} = 1$ and the same applied inequality between the arithmetical and the geometrical means of $n \geq 2$ positive real numbers, considering the triplet $ctg\hat{A}\cdot ctg\hat{B},\ \ ctg\hat{B}\cdot ctg\hat{C},\ \ ctg\hat{C}\cdot ctg\hat{A}$ generates the inequality $ctg\hat{A}\cdot ctg\hat{B}\cdot ctg\hat{C} \leq \dfrac{1}{3\sqrt{3}}$ (for the triangles with obtuse angles, the inequality is obvious).

As we have shown in trigonometry, the inequalities are the result of using the specific properties of the fundamental trigonometric functions correspondingly combined with adequate considerations from algebra, mathematical analysis, geometry etc.

Comments and Concluding Remarks

This book represents a novel approach for the trigonometry and a scientific work in this field by using the ensemble structure, the real analysis and the axiomatic fundaments of geometry (chapter 1-3). From chapter 4 to the end of this e-book one presents definitions, properties, formulae and applications more specific of the subject title. The book is recommanded as a pertinent introduction for the high school students, being also very useful for the university students. In a new edition, it will be completed adding new parts like Fourier series, hyperbolic trigonometry, geometric relations between circular and hyperbolic functions and so on.

REFERENCES

[1] Fraenkel A., *Abstract Set Theory.Amsterdam*, North Holland Publishing Company, 1953.

[2] Nachbin L., *Topology and Order,* New York, Van Nostrand, 1965.

[3] Peressini A.L., *Ordered Topological Vector Spaces,* Harper & Row, New York, 1967.

[4] Taylor A., *General Theory of Functions and Integration*, New York, Toronto, London, Blaisd. Publ. Com., 1965 .

[5] Postolică V., *Syntheses from The Foundations of Mathematics,* Matrix Rom Publishing House, Bucharest, Romania, 1999.

[6] Vaisman I., *Foundations of Mathematics*, Romanian Teaching and Pedagogical Publishing House, Bucharest, Romania, 1968.

Subject Index

A

Adherent 22, 23
Adherent points 21-23
Antinomies 4, 5
Applications of Mathematics 4
Arbitrary non-empty sets 5, 9, 10, 22, 52
Arbitrary non-void sets 8
Arbitrary sets, non-empty 8
Arbitrary straight line crosses 51
Arcsin arccos 94, 99, 104
Axiomatic construction 12, 55
Axiomatic introduction 3, 11
Axiomatic system, first complete 3, 49
Axiomatic theory 3
Axiomatized theories 3
Axiom of completeness Cantor-Dedekind 17
Axioms 1-3, 5-7, 14, 17, 18, 49, 53, 54

B

Bijective function 9, 10, 31, 73, 74
Binary relation 8-10
Bisectrix, first 65, 74, 80
Boundedness 35, 36, 66, 70
   lower 17, 18
   upper 17, 18

C

Calculus by identities 84
Cantor-Dedekind 1, 17, 18
Cardinal 1, 9, 12, 23, 28, 43, 74
Categoricalness 3, 4
Centers(axes) of symmetry 29
Circle, circumscribed 114, 115
Commutative field, ordered 17
Completeness axiom 1
Completeness Cantor-Dedekind 17, 18
Complex numbers 13, 84, 108-110
Composed function 10
Concave 29, 45, 47, 66, 70, 82, 122
Concave function 47, 48
Concavity 66, 70, 77, 82, 94, 95, 97, 123
Congruent 52, 53
Convergent 18, 20-22, 27
Convex 29, 45-47, 82, 96
Convex function 46, 47, 66, 70
Convexity 66, 70, 77, 82, 96, 97
Convex quadrilaterals 120, 121
Coordinates
   point of 71, 77
   system of 62, 63, 67, 71, 77
Cosine function 67-71, 77, 85, 94, 95, 118, 123
Cosines 60, 66, 70, 76, 77, 85, 118, 121
Cotangent function 60, 77-79, 81, 82, 97
Cotangent trigonometric function 66

D

Darboux property 29, 40-44, 66
Darboux's property 41
Decimals 12
Dedekind 15, 25, 54
Dedekind Principle 54
Dedekind section 54
Deduction 1
Definition set 26, 43, 77, 79, 82
Distinct points 50, 51
   arbitrary 51
Division property 52

E

Equality relation 9
Equilateral 122, 123
Equipolence 55
Equipolent 55, 56
Equivalence 9, 13-15, 52, 55, 56, 99
Euclidean geometry 2, 49, 55, 84
Euclidean's moment 49
Euclid's Postulate 1, 2

F

First rate discontinuity points 27
Function
   arbitrary 35
   bounded 27
   constant 35, 37
   contangent 89
   continuous 43, 66
   decreasing 26, 35
   discontinuos 44
   hyperbolic 125
   identical 30
   increasing 26, 35
   inverse 33, 34
   inverted 60
   module 40
   monotonous 36
   periodic 37
   random 31
   real 37
   unique 11
Functional relation 10
Function arccos 104
Function arcsin 65

G

Geometrical applications 113
Geometrical definition 63, 67, 71
Geometric characterization 29
Geometry 1-3, 49, 124